ASSESSMENT ITEM LISTING FOR FORCES, MOTION, AND ENERGY

HOLT, RINEHART AND WINSTON

A Harcourt Classroom Education Company

Austin · New York · Orlando · Atlanta · San Francisco · Boston · Dallas · Toronto · London

Art and Photo Credits

All work, unless otherwise noted, contributed by Holt, Rinehart and Winston.

Front cover: Jose Fuste Raga/The Stock Market; (owl on cover, title page) Kim Taylor/Bruce Coleman, Inc.

Printed in the United States of America

ISBN 0-03-065531-5

2 3 4 5 6 082 05 04 03 02

CONTENTS

ASSESSMENT ITEM LISTING

Introduction

The *Holt Science and Technology* Test Generator and *Assessment Item Listing*

The *Holt Science and Technology* Test Generator consists of a comprehensive bank of test items and the ExamView® Pro 3.0 software, which enables you to produce your own tests based on the items in the Test Generator and items you create yourself. Both Macintosh® and Windows® versions of the Test Generator are included on the *Holt Science and Technology* One-Stop Planner with Test Generator. Directions on pp. vi–vii of this book explain how to install the program on your computer. This *Assessment Item Listing* is a printout of all the test items in the *Holt Science and Technology* Test Generator.

ExamView Software

ExamView enables you to quickly create printed and on-line tests. You can enter your own questions in a variety of formats, including true/false, multiple choice, completion, problem, short answer, and essay. The program also allows you to customize the content and appearance of the tests you create.

Test Items

The *Holt Science and Technology* Test Generator contains a file of test items for each chapter of the textbook. Each item is correlated to the chapter objectives in the textbook and by difficulty level.

Item Codes

As you browse through this *Assessment Item Listing*, you will see that all test items of the same type appear under an identifying head. Each item is coded to assist you with item selection. Following is an explanation of the codes.

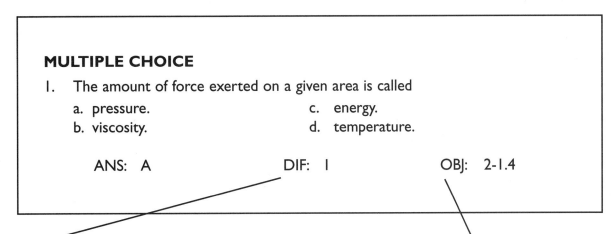

```
MULTIPLE CHOICE
  1.  The amount of force exerted on a given area is called
      a. pressure.                        c. energy.
      b. viscosity.                       d. temperature.

          ANS:  A              DIF:  I              OBJ:  2-1.4
```

DIF defines the difficulty of the item.
  I requires recall of information.
  II requires analysis and interpretation of known information.
  III requires application of knowledge to new situations.

OBJ lists the chapter number, section number, and objective.
(2-1.4 = Chapter 2, Section 1, Objective 4)

INSTALLATION AND STARTUP

The Test Generator is provided on the One-Stop Planner. The Test Generator includes ExamView and all of the questions for the corresponding textbook. ExamView includes three components: Test Builder, Question Bank Editor, and Test Player. The Test Builder includes options to create, edit, print, and save tests. The Question Bank Editor lets you create or edit question banks. The Test Player is a separate program that your students can use to take on-line* (computerized or LAN-based) tests. Please refer to the ExamView User's Guide on the One-Stop Planner for complete instructions.

Before you can use the Test Generator, you must install ExamView and the test banks on your hard drive. The system requirements, installation instructions, and startup procedures are provided below.

SYSTEM REQUIREMENTS

To use ExamView, your computer must meet or exceed the following hardware requirements:

Windows®
- Pentium processor
- Windows 95®, Windows 98®, Windows 2000® (or a more recent version)
- color monitor (VGA-compatible)
- CD-ROM and/or high-density floppy disk drive
- hard drive with at least 7 MB space available
- 8 MB available memory (16 MB memory recommended)
- an Internet connection (if you wish to access the Internet testing features)*

Macintosh®
- PowerPC processor, 100 MHz
- System 7.5 (or a more recent version)
- color monitor (VGA-compatible)
- CD-ROM and/or high-density floppy disk drive
- hard drive with at least 7 MB space available
- 8 MB available memory (16 MB memory recommended)
- an Internet connection with System 8.6 or a more recent version (if you wish to access the Internet testing features)*

* You can use the Test Player to host tests on your personal or school Web site or local area network (LAN) at no additional charge. The ExamView Web site's Internet test-hosting service must be purchased separately. Visit www.examview.com to learn more.

INSTALLATION

Instructions for installing ExamView from the CD-ROM:

Windows®
Step 1
Turn on your computer.
Step 2
Insert the One-Stop Planner into the CD-ROM drive.
Step 3
Click the Start button on the taskbar, and choose the Run option.
Step 4
In the Open box, type "d:\setup.exe" (substitute the letter for your drive if it is not d:) and click OK.
Step 5
Follow the prompts on the screen to complete the installation process.

Macintosh®
Step 1
Turn on your computer.
Step 2
Insert the One-Stop Planner into the CD-ROM drive. When the CD-ROM icon appears on the desktop, double-click the icon.
Step 3
Double-click the ExamView Pro Installer icon.
Step 4
Follow the prompts on the screen to complete the installation process.

Instructions for installing ExamView from the Main Menu of the One-Stop Planner (Macintosh® or Windows®):

Follow steps 1 and 2 from above.
Step 3
Double-click One-Stop.pdf. (If you do not have Adobe Acrobat® Reader installed on your computer, install it before proceeding by clicking Reader Installer.)
Step 4
To advance to the Main Menu, click anywhere on the title screen.
Step 5
Click the Test Generator button.
Step 6
Click the appropriate Install ExamView button.
Step 7
Follow the prompts on the screen to complete the installation process.

GETTING STARTED

After you complete the installation process, follow these instructions to start ExamView. See the ExamView User's Guide on the One-Stop Planner for further instructions on the program's options for creating a test and editing a question bank.

Startup Instructions

Step 1
Turn on the computer.
Step 2
Windows®: Click the Start button on the taskbar. Highlight the Programs menu, and locate the ExamView Pro Test Generator folder. Select the ExamView Pro option to start the software.
Macintosh®: Locate and open the ExamView Pro folder. Double-click the ExamView Pro icon.
Step 3
The first time you run the software, you will be prompted to enter your name, school/institution name, and city/state. You are now ready to begin using ExamView.
Step 4
Each time you start ExamView, the Startup menu appears. Choose one of the options shown.
Step 5
Use ExamView to create a test or edit questions in a question bank.

Technical Support

If you have any questions about the Test Generator, call the Holt, Rinehart and Winston technical support line at 1-800-323-9239, Monday through Friday, 7:00 A.M. to 6:00 P.M., Central Standard Time. You can contact the Technical Support Center on the Internet at http://www.hrwtechsupport.com or by e-mail at tsc@hrwtechsupport.com.

MULTIPLE CHOICE

1. A student riding her bicycle on a straight, flat road covers one block every 7 seconds. If each block is 100 m long, she is traveling at
   a. constant speed.
   b. constant velocity.
   c. 10 m/s.
   d. Both (a) and (b)

   ANS: D   DIF: II   OBJ: 1-1.3

2. Friction is a force that
   a. opposes an object's motion.
   b. does not exist when surfaces are very smooth.
   c. decreases with larger mass.
   d. All of the above

   ANS: A   DIF: I   OBJ: 1-3.1

3. Rolling friction
   a. is usually less than sliding friction.
   b. makes it difficult to move objects on wheels.
   c. is usually greater than sliding friction.
   d. is the same as fluid friction.

   ANS: A   DIF: I   OBJ: 1-3.2

4. If Earth's mass doubled, your weight would
   a. increase because gravity increases.
   b. decrease because gravity increases.
   c. increase because gravity decreases.
   d. not change because you are still on Earth.

   ANS: A   DIF: I   OBJ: 1-4.1

5. A force
   a. is expressed in newtons.
   b. can cause an object to speed up, slow down, or change direction.
   c. is a push or a pull.
   d. All of the above

   ANS: D   DIF: I   OBJ: 1-2.1

6. The amount of gravity between 1 kg of lead and Earth is _____ the amount of gravity between 1 kg of marshmallows and Earth.
   a. greater than
   b. less than
   c. the same as
   d. None of the above

   ANS: C   DIF: I   OBJ: 1-4.2

7. If passengers on an airplane watch another plane passing by them, their frame of reference is the
   a. sky.
   b. ground.
   c. other plane.
   d. plane they are on.

   ANS: D   DIF: I   OBJ: 1-1.1

8. Which type of friction causes a sky diver's acceleration to change as she falls?
   a. sliding
   b. rolling
   c. fluid
   d. static

   ANS: C   DIF: I   OBJ: 1-3.2

9. To calculate an object's acceleration, you need to know
   a. distance traveled and total time.
   b. starting point, endpoint, and the object's mass.
   c. starting velocity, final velocity, and time it takes to change velocity.
   d. average speed and direction traveled.

   ANS: C   DIF: I   OBJ: 1-1.4

10. The force of gravity is greater between two objects that
    a. have greater masses.
    b. have rougher surfaces.
    c. are farther apart.
    d. are moving at greater speed.

    ANS: A   DIF: I   OBJ: 1-4.1

11. Two forces act on an object. One force has a magnitude of 10 N and is directed toward the north. The other has a magnitude of 5 N directed toward the south. The object experiences a net force of
    a. 5 N south.
    b. 15 N north.
    c. 50 N north.
    d. 5 N north.

    ANS: D   DIF: II   OBJ: 1-2.2

12. A reference point for determining position and motion could be
    a. the Earth's surface.
    b. a building.
    c. a moving object.
    d. All of the above

    ANS: D   DIF: I   OBJ: 1-1.1

13. The distance traveled divided by the time it took to travel that distance determines an object's
    a. speed.
    b. acceleration.
    c. weight.
    d. force.

    ANS: A   DIF: I   OBJ: 1-1.2

14. The SI unit for speed is
    a. km/h.
    b. f/s.
    c. m/s.
    d. m/h.

    ANS: C   DIF: I   OBJ: 1-1.2

15. In 1.5 h you walk 7.5 km. When you divide 7.5 km by 1.5 h, you are calculating your
    a. instantaneous speed.
    b. average speed.
    c. velocity.
    d. acceleration.

    ANS: B   DIF: I   OBJ: 1-1.2

16. A bird flies 150 m for 10 s, then 200 m for 10 s, and then 100 m for 5 s. What is the bird's average speed?
    a. 150 m/s
    b. 100 m/s
    c. 18 m/s
    d. 8.33 m/s

    ANS: C   DIF: II   OBJ: 1-1.2

17. You are in Chicago, IL. You decide to head south to Austin, TX. In one hour, you travel 80 km. Your velocity is
    a. 80:1.
    b. 80 km.
    c. 80 km/h.
    d. 80 km/h south.

    ANS: D   DIF: II   OBJ: 1-1.3

18. If a bus traveling 15 m/s south speeds up to 20 m/s, this is a change in its
    a. speed.
    b. velocity.
    c. acceleration.
    d. All of the above

    ANS: D   DIF: I   OBJ: 1-1.4

19. A car traveling 20 m/s south enters a new highway going east at 20 m/s. The car has
    a. decreased its distance.
    b. changed its speed.
    c. accelerated.
    d. not changed its velocity.

    ANS: C   DIF: II   OBJ: 1-1.4

20. You are on a bus traveling 15 m/s east and you decide to move from the front of the bus to the back walking at a rate of 1 m/s. Your resultant velocity is
    a. 1 m/s west.
    b. 15 m/s east.
    c. 14 m/s east.
    d. 14 m/s west.

    ANS: C   DIF: II   OBJ: 1-1.3

21. The space shuttle is always launched in the same direction that the Earth rotates. This allows the shuttle to use less fuel because it takes advantage of the Earth's
    a. circular speed.
    b. rotational velocity.
    c. acceleration.
    d. gravity.

    ANS: B   DIF: I   OBJ: 1-1.3

22. Acceleration is a change in
    a. speed.
    b. velocity.
    c. direction.
    d. All of the above

    ANS: D   DIF: I   OBJ: 1-1.4

23. A plane passes over Somton with a velocity of 8,000 m/s north. The plane passes over Allton at a velocity of 10,000 m/s north 40 s later. What is the plane's acceleration from Somton to Allton?
   a. 2,000 m/s north.
   b. 2,000 m/s/s north.
   c. 50 m/s/s north.
   d. 50 m/s north.

   ANS: C   DIF: II   OBJ: 1-1.4

24. A cheetah runs eastward at a velocity of 27 m/s. Two seconds later, it tackles its prey to the ground. What is the cheetah's acceleration?
   a. 27 m/s eastward
   b. 27 m/s/s eastward
   c. 13.5 m/s eastward
   d. −13.5 m/s/s eastward

   ANS: D   DIF: II   OBJ: 1-1.4

25. When velocity decreases, this could be referred to as
   a. acceleration.
   b. deceleration.
   c. negative acceleration.
   d. All of the above

   ANS: D   DIF: I   OBJ: 1-1.4

26. An example of acceleration is
   a. a hovering helicopter.
   b. a car turning a corner.
   c. jogging at the same pace.
   d. bicycling at 40 m/s.

   ANS: B   DIF: I   OBJ: 1-1.4

27. The moving blades of a windmill are an example of
   a. centripetal acceleration.
   b. positive acceleration.
   c. negative acceleration.
   d. deceleration.

   ANS: A   DIF: I   OBJ: 1-1.4

28. Which of the following is NOT an example of a force being exerted?
   a. pushing a door open
   b. typing on a computer keyboard
   c. sitting in a chair
   d. none of the above

   ANS: D   DIF: I   OBJ: 1-2.1

29. A friend is helping you to rearrange your bedroom furniture. You exert 25 N of force to push the foot-end of the bed while your friend exerts 20 N of force pulling the head-end of the bed. The net force being used to move the bed is
   a. 5 N forward.
   b. 20 N forward.
   c. 25 N forward.
   d. 45 N forward.

   ANS: D   DIF: II   OBJ: 1-2.2

30. What is the net force when you combine a force of 7 N north with a force of 5 N south?
   a. 2 N north
   b. 2 N south
   c. 12 N north
   d. 12 N south

   ANS: A   DIF: II   OBJ: 1-2.2

31. Balanced forces applied to an object
    a. produce a net force of zero.
    b. change the direction of a moving object.
    c. cause an object at rest to start moving.
    d. produce a negative net force.

    ANS: A   DIF: I   OBJ: 1-2.3

32. A force that opposes motion between two surfaces that are touching is
    a. gravity.
    b. friction.
    c. velocity.
    d. acceleration.

    ANS: B   DIF: I   OBJ: 1-3.1

33. Every year, Cub Scouts hold a car derby in which the scouts create their own car from a block of wood. Each car is weighed before the race and then raced on identical tracks. To ensure a fair race, all cars should weigh close to the same amount because
    a. a less massive car exerts less force on the track, creates less friction, and goes faster.
    b. a more massive car exerts less force on the track, creates less friction, and goes faster.
    c. a less massive car exerts less force on the track, creates more friction, and goes faster.
    d. a more massive car exerts less force on the track, creates more friction, and goes faster.

    ANS: A   DIF: I   OBJ: 1-3.1

34. A good rule to observe when around a swimming pool is "no running." The deck around the pool becomes very slippery to walk on when wet due to
    a. sliding friction.
    b. rolling friction.
    c. fluid friction.
    d. static friction.

    ANS: C   DIF: I   OBJ: 1-3.2

35. One of the largest forces opposing the motion of a car is
    a. sliding friction.
    b. rolling friction.
    c. fluid friction.
    d. static friction.

    ANS: C   DIF: I   OBJ: 1-3.2

36. One way to increase friction is to use
    a. wax.
    b. water.
    c. sand.
    d. oil.

    ANS: C   DIF: I   OBJ: 1-3.3

37. After several thousand kilometers, the treads on a car's tires wear away, making them very smooth. Which type of friction is mainly responsible for this?
    a. sliding
    b. rolling
    c. fluid
    d. static

    ANS: B   DIF: I   OBJ: 1-3.2

38. The gravitational force of the Earth in relation to the gravitational force of everything else on Earth represents a(n) ____ that causes everything to fall toward the center of the Earth.
   a. unbalanced force
   b. balanced force
   c. centrifugal force
   d. centripetal force

   ANS: A   DIF: I   OBJ: 1-4.1

39. The law of universal gravitation describes the relationships between
   a. speed, velocity, and acceleration.
   b. unbalanced force, balanced force, and net force.
   c. speed, distance, and time.
   d. force, mass, and distance.

   ANS: D   DIF: I   OBJ: 1-4.2

40. Objects of any size exert a ____ force.
   a. gravitational
   b. balanced
   c. centrifugal
   d. centripetal

   ANS: A   DIF: I   OBJ: 1-4.2

41. The planets are held in their orbits by ____ forces.
   a. balanced
   b. unbalanced
   c. centrifugal
   d. centripetal

   ANS: B   DIF: I   OBJ: 1-4.2

42. Gravitational force depends on the ____ of the objects and the distance between them.
   a. weight
   b. mass
   c. speed
   d. force

   ANS: B   DIF: I   OBJ: 1-4.2

43. A black hole's gravitational force is incredibly large due to its
   a. mass.
   b. color.
   c. radius.
   d. distance.

   ANS: A   DIF: I   OBJ: 1-4.2

44. You will drive to a town that is 100 km away. Half the trip you will be able to drive 100 km/h (on the highway) and half the trip you will drive 50 km/h (going through small towns). How long will it take you to get there?
   a. 30 min
   b. 1.1 h
   c. 1.5 h
   d. 2 h

   ANS: C   DIF: II   OBJ: 1-1.2

45. You are training to run the 100 m race at a track and field meet. You time yourself and find you can run it in 10 seconds. How fast are you able to run it?
   a. 1 m/s
   b. 10 m/s
   c. 100 m/s
   d. 1,000 m/s

   ANS: B   DIF: II   OBJ: 1-1.2

46. In an action film, the hero is running forward along the top of a train going 40 m/s. He's chasing the villain at a rate of 3 m/s. The hero's resultant velocity is
    a. 3 m/s forward.
    b. 37 m/s forward.
    c. 40 m/s forward.
    d. 43 m/s forward.

    ANS: D   DIF: II   OBJ: 1-1.3

47. You ride your bike north for 100 m at a constant speed of 5 m/s. Your acceleration is
    a. 5 m/s/s forward.
    b. 20 m/s/s forward.
    c. 100 m/s/s forward.
    d. There is no acceleration.

    ANS: D   DIF: II   OBJ: 1-1.4

48. Gravitational force increases as
    a. acceleration increases.
    b. mass increases.
    c. distance increases.
    d. velocity increases.

    ANS: B   DIF: I   OBJ: 1-4.2

49. Mass is measured with a
    a. hydrometer.
    b. balance.
    c. newtonmeter.
    d. barometer.

    ANS: B   DIF: I   OBJ: 1-4.3

50. The air that comes out of the tiny holes of an air-hockey table is an example of
    a. net force.
    b. friction.
    c. gas lubricant.
    d. liquid lubricant.

    ANS: C   DIF: I   OBJ: 1-3.3

51. A hot air balloon floats low in the sky. You see it pass your house and 10 seconds later pass a building you know to be 50 m away. You estimate the speed of the balloon to be
    a. 0.2 m/s.
    b. 5 m/s.
    c. 10 m/s.
    d. 50 m/s.

    ANS: B   DIF: I   OBJ: 1-1.2

52. An object's velocity remains constant when its ____ do not change.
    a. speed and direction
    b. speed and time in motion
    c. time in motion and direction
    d. distance traveled and time in motion

    ANS: A   DIF: I   OBJ: 1-1.3

53. An air-traffic controller is told that an airplane's speed is 600 km/h and that its destination is 1200 km south of its present position. The controller is asked to calculate how long it will take the airplane to reach its destination. The controller is unable to do this calculation because the information given was incomplete. It makes a difference as to what the airplane's ____ is.
    a. altitude
    b. passenger load
    c. direction
    d. gas mileage

    ANS: C   DIF: I   OBJ: 1-1.3

54. At the bottom of a hill, a runner jogs 4 m/s north. One minute later—at the top of the hill—the jogger's velocity is only 1 m/s north. What is the jogger's acceleration from the bottom to the top of the hill?

a. 3 m/s/s north

c. $\frac{1}{20}$ m/s/s north

b. −3 m/s/s north

d. $-\frac{1}{20}$ m/s/s north

ANS: D   DIF: II   OBJ: 1-1.4

55. In a storm, the wind blows with a velocity of 10 m/s south. Five seconds later, the wind's velocity is 25 m/s south. What is the acceleration of the wind?

a. 3 m/s/s south   c. 15 m/s/s south
b. 6 m/s/s south   d. 25 m/s/s south

ANS: A   DIF: II   OBJ: 1-1.4

56. You ride your bike around the block at a constant 11 km/h. What changes?

a. your velocity   c. Both (a) and (b)
b. your acceleration   d. Neither (a) nor (b)

ANS: C   DIF: I   OBJ: 1-1.4

57. You are in-line skating on a path. As you skate up a slight hill, you slow down. This illustrates

a. acceleration.   c. deceleration.
b. negative acceleration.   d. All of the above

ANS: D   DIF: I   OBJ: 1-1.4

58. A yellow ball rolls at a constant velocity while a red ball accelerates. The red ball may be

a. moving faster than the yellow ball.   c. curving away from the yellow ball.
b. slowing down.   d. Any of the above

ANS: D   DIF: I   OBJ: 1-1.4

59. Your class and another class have a tug-of-war contest. For your class to win, your class must

a. create a balanced force with the other class.
b. have a greater force in the direction opposite the other class.
c. have a greater force in the same direction as the other class.
d. create a net force of zero with the other class.

ANS: B   DIF: I   OBJ: 1-2.3

60. For motion to occur,

a. a net force must be greater than zero.   c. acceleration must be created.
b. an unbalanced force must be exerted.   d. All of the above

ANS: D   DIF: I   OBJ: 1-2.3

61. You have made a house of cards on top of your table. Suddenly, a gust of wind blows through an open window and your house of cards comes tumbling down. The wind applied _____ to your house of cards.
    a. friction
    b. a balanced force
    c. an unbalanced force
    d. a gravitational force

    ANS: C   DIF: I   OBJ: 1-2.3

62. When applying brakes on a car or a bicycle, you are using _____ friction to stop.
    a. sliding
    b. rolling
    c. fluid
    d. static

    ANS: A   DIF: I   OBJ: 1-3.2

63. One Saturday, you go on a picnic. Although there is a slight breeze, the napkins stay on the tablecloth because of
    a. sliding friction.
    b. rolling friction.
    c. fluid friction.
    d. static friction.

    ANS: D   DIF: I   OBJ: 1-3.2

64. An autumn breeze blows leaves from the trees. Although the leaves are flying through the air, they are still experiencing
    a. sliding friction.
    b. rolling friction.
    c. fluid friction.
    d. static friction.

    ANS: C   DIF: I   OBJ: 1-3.2

65. Rolling friction is usually _____ sliding friction.
    a. less than
    b. equal to
    c. slightly greater than
    d. quite a bit greater than

    ANS: A   DIF: I   OBJ: 1-3.2

66. Venus and Earth have nearly the same mass. Therefore, the gravitational pull from the sun is approximately
    a. the same for both even though Venus is closer to the sun.
    b. greater for Venus because it is closer to the sun.
    c. less for Venus because it is closer to the sun.
    d. greater for Earth because it is farther from the sun.

    ANS: B   DIF: I   OBJ: 1-4.2

67. When gravitational force changes,
    a. mass changes.
    b. weight changes.
    c. the weight stays the same.
    d. direction always changes.

    ANS: B   DIF: I   OBJ: 1-4.3

68. Because the moon's gravitational force is one-sixth that of Earth's, an astronaut's weight on the moon is
   a. the same as on Earth because gravitational force affects mass, not weight.
   b. six times that on Earth.
   c. one-sixth that on Earth.
   d. one-third that on Earth because Earth's gravity is also pulling on the astronaut.

   ANS: C   DIF: I   OBJ: 1-4.3

COMPLETION

1. _____ opposes motion between surfaces that are touching. (Friction or Gravity)

   ANS: Friction   DIF: I   OBJ: 1-3.1

2. Forces are expressed in _____. (newtons or mass)

   ANS: newtons   DIF: I   OBJ: 1-2.1

3. A _____ is determined by combining forces. (net force or newton)

   ANS: net force   DIF: I   OBJ: 1-2.2

4. _____ is the rate at which _____ changes. (Velocity or Acceleration, velocity or acceleration)

   ANS: Acceleration, velocity   DIF: I   OBJ: 1-1.4

5. Any change in position over time is an example of _____. (motion or acceleration)

   ANS: motion   DIF: I   OBJ: 1-1.1

6. The newton is the SI unit of _____. (force or mass)

   ANS: force   DIF: I   OBJ: 1-2.1

7. A change in a moving object's direction always results in a change in its _____. (speed or velocity)

   ANS: velocity   DIF: I   OBJ: 1-1.3

8. A _____ force always acts to oppose motion. (frictional or gravitational)

   ANS: frictional   DIF: I   OBJ: 1-3.1

9. The _____ of an object can change with its location. (mass or weight)

   ANS: weight   DIF: I   OBJ: 1-4.1

10. _____ is the rate at which an object moves, but it does not include direction.

    ANS: Speed   DIF: I   OBJ: 1-1.2

11. When observing an object in motion in relation to an object that appears to stay in place, the object that appears to stay in place is a _____.

    ANS: reference point   DIF: I   OBJ: 1-1.1

12. Total distance divided by total time gives an object's average _____.

    ANS: speed   DIF: I   OBJ: 1-1.2

13. The speed of an object in a particular direction is an object's _____.

    ANS: velocity   DIF: I   OBJ: 1-1.3

14. The acceleration that occurs in circular motion is known as _____ acceleration.

    ANS: centripetal   DIF: I   OBJ: 1-1.4

15. A push or a pull is a _____.

    ANS: force   DIF: I   OBJ: 1-2.1

16. When the net force on an object is not zero, the forces on the object are _____.

    ANS: unbalanced   DIF: I   OBJ: 1-2.3

17. When a force is applied to an object but does NOT cause the object to move, _____ occurs.

    ANS: static friction   DIF: I   OBJ: 1-3.2

18. _____ are substances that are applied to surfaces in order to reduce the friction between them.

    ANS: Lubricants   DIF: I   OBJ: 1-3.3

19. _____ is the force of attraction between objects that is due to their masses.

    ANS: Gravity   DIF: I   OBJ: 1-4.1

20. _____ is a measure of the gravitational force of an object.

    ANS: Weight   DIF: I   OBJ: 1-4.3

21. _____ is the amount of matter in an object.

    ANS: Mass   DIF: I   OBJ: 1-1.1

22. Most of the time, objects do not travel at _____ speed, which is why it is useful to calculate an object's average speed.

    ANS: constant   DIF: I   OBJ: 1-1.2

23. Cars traveling in opposite directions on a highway may attain the same speed, but never the same _____.

    ANS: velocity   DIF: I   OBJ: 1-1.3

24. Sir Isaac Newton's discoveries led to the law of _____.

    ANS: universal gravitation   DIF: I   OBJ: 1-4.2

SHORT ANSWER

1. What is a reference point?

   ANS:
   A reference point is an object that appears to stay in place in relation to an object being observed and is used to determine if the object is in motion.

   DIF: I   OBJ: 1-1.1

2. What two things must you know to determine speed?

   ANS:
   You must know the distance traveled and the time taken to travel that distance.

   DIF: I   OBJ: 1-1.2

3. What is the difference between speed and velocity?

   ANS:
   Speed does not include direction; velocity does.

   DIF: I   OBJ: 1-1.3

4. Explain why it is important to know a tornado's velocity and not just its speed.

ANS:
It would be important to know the velocity because velocity includes direction. Knowing only the speed of a tornado would not tell the direction that the tornado is traveling. Knowing a tornado's direction of travel would allow people to avoid or escape its path.

DIF: II   OBJ: 1-1.3

5. What is acceleration?

ANS:
Acceleration is the rate at which velocity changes.

DIF: I   OBJ: 1-1.4

6. Does a change in direction affect acceleration? Explain your answer.

ANS:
Yes, a change in direction does affect acceleration. Acceleration is a measure of velocity change. Velocity is speed in a given direction, and a velocity changes if direction changes.

DIF: I   OBJ: 1-1.4

7. How do you think a graph of deceleration would differ from the graph shown below? Explain your reasoning.

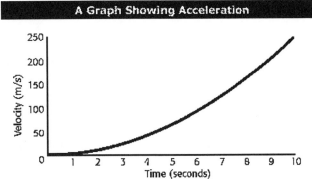

ANS:
The graph showing acceleration has a positive slope. A graph showing deceleration would have a negative slope. The graph would take this shape because velocity would be decreasing as time passes.

DIF: II   OBJ: 1-1.5

8. Give four examples of a force being exerted.

    ANS:
    Accept all reasonable answers. Examples include: kicking a ball, writing with a pencil, pulling a rope, and pushing a stalled car.

    DIF: I   OBJ: 1-2.1

9. Explain the difference between balanced and unbalanced forces and how each affects the motion of an object.

    ANS:
    Unbalanced forces occur when the net force on an object is not zero and cause a change in an object's motion. *Balanced forces* occur when the net force equals zero and do not cause a change in an object's motion.

    DIF: I   OBJ: 1-2.3

10. Explain why friction occurs.

    ANS:
    Friction occurs because the microscopic hills and valleys of two touching surfaces "stick" to each other.

    DIF: I   OBJ: 1-3.1

11. Name two ways in which friction can be increased.

    ANS:
    Friction can be increased by making surfaces rougher and by increasing the force pushing the surfaces together.

    DIF: I   OBJ: 1-3.1

12. Give an example of each of the following types of friction: sliding, rolling, and fluid.

    ANS:
    Answers will vary; accept all reasonable answers. Sample answers: *Sliding* friction is used while skiing and writing with a pencil. *Rolling* friction is used while riding a bicycle and pushing a handcart. *Fluid* friction is used while swimming and throwing a softball.

    DIF: I   OBJ: 1-3.2

13. Name two ways that friction is harmful and two ways that friction is helpful to you when riding a bicycle.

ANS:
Accept all reasonable answers. Sample answer: Friction can be *harmful* because it causes tire tread to wear down and the wind can slow you down. Friction can be *helpful* because the wheels grip the road, and your feet and hands stay on the pedals and handlebars.

DIF: II   OBJ: 1-3.3

14. How does the mass of an object relate to the gravitational force the object exerts on other objects?

ANS:
The greater an object's mass, the larger the gravitational force it exerts on other objects.

DIF: I   OBJ: 1-4.2

15. How does the distance between objects affect the gravity between them?

ANS:
As the distance between objects increases, the gravitational force between them decreases; as the distance between objects decreases, the gravitational force between them increases.

DIF: I   OBJ: 1-4.2

16. Explain why your weight would change if you orbited Earth in the space shuttle but your mass would not.

ANS:
A person's *weight* decreases in orbit because the distance between the person and the Earth would increase. But the person's *mass* would remain constant, because mass is the amount of matter in an object, and it does not depend on gravitational force.

DIF: II   OBJ: 1-4.3

17. What distinguishes the measurement of speed from that of velocity and acceleration?

ANS:
Speed does not involve direction, as both *velocity* and *acceleration* do.

DIF: I   OBJ: 1-1.3

18. What is centripetal acceleration?

ANS:
Centripetal acceleration is acceleration that occurs in circular motion.

DIF: I   OBJ: 1-1.5

19. How do you calculate speed?

   ANS:
   To calculate speed, divide the distance traveled by the time.

   DIF: I   OBJ: 1-1.2

20. How do you calculate velocity?

   ANS:
   To calculate velocity, divide the distance and direction traveled by the time.

   DIF: I   OBJ: 1-1.3

21. How do you calculate acceleration?

   ANS:
   To calculate acceleration, subtract the starting velocity from the final velocity, and divide it by the time it takes to change velocity.

   DIF: I   OBJ: 1-1.4

22. What is a net force?

   ANS:
   A net force is the sum of all the forces acting on an object.

   DIF: I   OBJ: 1-2.2

23. Are the forces on a kicked soccer ball balanced or unbalanced? How do you know?

   ANS:
   Unbalanced; because the ball changes speed and/or direction.

   DIF: I   OBJ: 1-2.3

24. Which of the following would NOT help you move a heavy object across a concrete floor: water, ball bearings, oil, soapsuds, steel rods, foam rubber

   ANS:
   Foam rubber would not help move a heavy object across a concrete floor.

   DIF: I   OBJ: 1-3.3

25. Name three common items you might use to increase friction.

   ANS:
   Possible answers: sticky tape, sand, work gloves

   DIF: I   OBJ: 1-3.2

26. Name three common items you might use to reduce friction.

ANS:
Possible answers: oil, water, wax, grease

DIF: I   OBJ: 1-3.2

27. What is gravity?

ANS:
Gravity is a force of attraction between objects that is due to the masses of the objects.

DIF: I   OBJ: 1-4.3

28. What must you know in order to determine the gravitational force between two objects?

ANS:
To determine the gravitational force between two objects, you must know their masses and the distance between them.

DIF: I   OBJ: 1-4.2

29. Where would you weigh the most, on a boat, on the space shuttle, or on the moon?

ANS:
I would weigh the most on a boat.

DIF: I   OBJ: 1-4.1

30. Describe the relationship between motion and a reference point.

ANS:
Motion occurs when an object changes position over time when compared with a *reference point* (an object that appears to stay in place).

DIF: I   OBJ: 1-1.1

31. How is it possible to be accelerating and traveling at a constant speed?

ANS:
Acceleration can occur simply by a change in direction. Thus, no change in speed is necessary for acceleration.

DIF: II   OBJ: 1-1.4

32. Explain the difference between mass and weight.

ANS:
Mass is the amount of matter in an object, and its value does not change with the object's location. *Weight* is a measure of the gravitational force on an object, so an object's weight can change as the amount of gravitational force changes.

DIF: I   OBJ: 1-4.3

33. Use the following terms to create a concept map: *speed, velocity, acceleration, force, direction, motion.*

ANS:

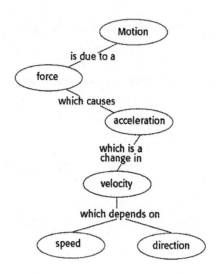

DIF: II   OBJ: 1-2.1

34. Your family is moving, and you are asked to help move some boxes. One box is so heavy that you must push it across the room rather than lift it. What are some ways you could reduce friction to make moving the box easier?

ANS:
Accept all reasonable responses. Examples include using a hand-cart or dolly to take advantage of rolling friction and polishing the floor to reduce sliding friction.

DIF: II   OBJ: 1-3.3

35. Explain how using the term accelerator when talking about a car's gas pedal can lead to confusion, considering the scientific meaning of the word acceleration.

ANS:
The car's gas pedal (*accelerator*) is pressed by the driver to increase the car's velocity. Since the scientific meaning of the term *acceleration* can include slowing down and even changing direction, accelerator is not an accurate term for this device.

DIF: II   OBJ: 1-1.4

36. Explain why it is important for airplane pilots to know wind velocity, not just wind speed, during a flight.

ANS:
It is helpful for pilots to know wind velocity because velocity includes direction. Pilots need to know the wind's speed and direction so that they will know whether the wind is blowing in the same direction as the plane (which could increase the plane's resultant velocity and lead to an earlier arrival time) or in a different direction than the plane (which might lead to a later arrival).

DIF: II   OBJ: 1-1.3

37. A kangaroo hops 60 m to the east in 5 s.
    a. What is the kangaroo's speed?
    b. What is the kangaroo's velocity?
    c. The kangaroo stops at a lake for a drink of water, then starts hopping again to the south. Every second, the kangaroo's velocity increases 2.5 m/s. What is the kangaroo's acceleration after 5 seconds?

ANS:
a.   12 m/s
b.   12 m/s east
c.   2.5 m/s/s south

DIF: II   OBJ: 1-1.4

38. Is this a graph of positive or negative acceleration? How can you tell?

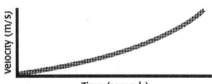

ANS:
Positive; because velocity increases as time passes

DIF: II   OBJ: 1-1.5

39. You know how to combine two forces that act in one or two directions. The same method you learned can be used to combine several forces acting in several directions. For each diagram below, predict with how much force and in what direction the object will move.

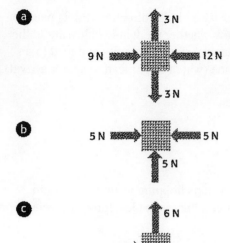

ANS:
a. 3 N to the left
b. 5 N up
c. 4 N to the right

DIF: II   OBJ: 1-2.3

40. What is the net force when you combine a force of 10 N north with a force of 2 N south?

ANS:
8 N north

DIF: I   OBJ: 1-4.3

The following graph shows the distance Kyle traveled on his bicycle trip and the amount of time it took him to travel that distance. Examine the graph and answer the following questions.

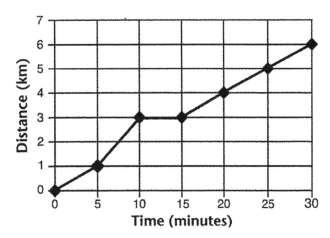

41. During which 5-minute interval did Kyle achieve the greatest average speed?

ANS:
Kyle achieved the greatest average speed during the interval between 5 minutes and 10 minutes. The distance traveled during this interval (2 km) was greater than the distance covered in any other 5-minute interval.

DIF: II   , OBJ: 1-1.5

42. What may have happened between $t = 10$ and $t = 15$ minutes?

ANS:
Kyle probably stopped. He did not travel any distance in this interval.

DIF: II   OBJ: 1-1.5

43. The design of a car can make a big difference in its maximum speed. Explain why two cars that have the same mass and are powered by the same type of engine might be able to reach very different speeds on the same road.

ANS:
Sample answer: As a car moves, air resistance (or fluid friction) works to slow the car's motion. A car that is designed to reduce this resistance will be able to reach higher speeds than a car that is not designed that way.

DIF: II   OBJ: 1-3.3

44. Use the following terms to complete the concept map below: *distance, newtons, balance, spring scale, attractive force, mass.*

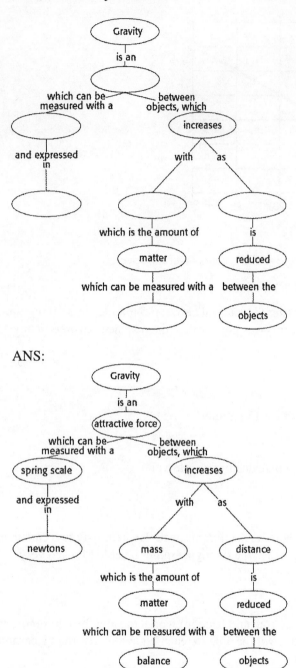

ANS:

DIF: II   OBJ: 1-4.3

45. A river is flowing at a velocity of 0.15 m/s west. If a swimmer is swimming against the current at a velocity of 0.22 m/s east, how many hours will it take him to swim 1 km? Show your work.

ANS:
Net velocity = 0.22 –0.15 = 0.07 m/s east
1,000 m ÷ 0.07 m/s = 14,285.71 s ÷ 3,600 s/h ≅ 4 h

DIF: II   OBJ: 1-1.4

46. You know how to combine two forces that act in one or two directions. The same method you learned can be used to combine several directions. Examine the diagram below and predict how much force and in what direction the object will move.

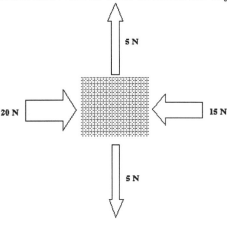

ANS:
The object will move 5 N to the right.

DIF: III   OBJ: 1-2.3

ESSAY

1. Furniture movers sometimes use a broom to help move heavy objects such as washing machines. One person can move a washing machine fairly easily by resting one of its bottom edges on the straw of the broom. The washer can then slide along the floor. Explain what is happening and why this works.

ANS:
There is less friction between the broom and the floor surface than there is between the bottom of the washing machine and the floor. The broom reduces the sliding friction between the washing machine and the floor.

DIF: II   OBJ: 1-3.3

2. Use the terms *forces* and *acceleration* to explain why you and the cars on a roller coaster do not fall off when you loop-the-loop.

ANS:
Accept all reasonable answers. Sample answer: An unbalanced force—the motor and chain that pulls the cars to the top of the first hill—produces a change in the motion of the cars, creating acceleration. Then the force of gravity acts on the cars, accelerating them downward. The cars speed up enough to overcome the forces of gravity and friction resulting in positive acceleration as they go around the loop-the-loop. Also, centripetal acceleration and both positive and negative acceleration all combine enough to overcome the frictional and gravitational forces exerted on both you and the cars and, thus, keep you and the cars moving along the direction of the track.

DIF: II   OBJ: 1-1.4

3. Use the terms *unbalanced forces* and the *law of universal gravitation* to explain how Sir Isaac Newton related why objects fall toward the Earth and what keeps the planets in orbit.

ANS:
Accept all reasonable answers. Sample answer: One night, Newton observed a falling apple and, realizing that unbalanced forces are necessary to move or change the motion of an object, he concluded that there must be an unbalanced force on the apple to make it fall and an unbalanced force on the moon to keep it moving in a circle around the Earth. He then realized that the force pulling the apple to the Earth and the force pulling the moon in a circular path around the Earth are the same force—a force of attraction called gravity. Newton generalized his observations on gravity in a law now known as the *law of universal gravitation,* which applies to all objects in the universe.

DIF: II   OBJ: 1-4.2

MULTIPLE CHOICE

1. A feather and a rock dropped at the same time from the same height would land at the same time when dropped
   a. by Galileo in Italy
   b. by Newton in England.
   c. by an astronaut on the moon.
   d. by an astronaut on the space shuttle.

   ANS: C   DIF: I   OBJ: 2-1.1

2. An object is in projectile motion if
   a. it is thrown with a horizontal push.
   b. it is accelerated downward by gravity.
   c. it does not accelerate horizontally.
   d. All of the above

   ANS: D   DIF: I   OBJ: 2-1.4

3. When a soccer ball is kicked, the action and reaction forces do NOT cancel each other out because
   a. the force of the foot on the ball is bigger than the force of the ball on the foot.
   b. the forces act on two different objects.
   c. the forces act at different times.
   d. All of the above

   ANS: B   DIF: I   OBJ: 2-2.1

4. Newton's first law of motion applies
   a. to moving objects.
   b. to objects that are not moving.
   c. to objects that are accelerating.
   d. Both (a) and (b)

   ANS: D   DIF: I   OBJ: 2-2.1

5. Acceleration of an object
   a. decreases as the mass of the object increases.
   b. increases as the force on the object increases.
   c. is in the same direction as the force on the object.
   d. All of the above

   ANS: D   DIF: I   OBJ: 2-2.1

6. A golf ball and a bowling ball are moving at the same velocity. Which has more momentum?
   a. the golf ball, because it has less mass
   b. the bowling ball, because it has more mass
   c. They both have the same momentum because they have the same velocity.
   d. There is no way to know without additional information.

   ANS: B   DIF: I   OBJ: 2-2.2

7. If three balls of different materials were dropped at the same time from the same height, which would hit the ground first? (Assume there is no air resistance.)
   a. a tennis ball
   b. a solid rubber ball
   c. a solid steel ball
   d. They would all hit at the same time.

   ANS: D   DIF: I   OBJ: 2-1.1

8. Orbital motion is a combination of forward motion and
   a. frictional resistance.
   b. free fall.
   c. horizontal acceleration.
   d. weightlessness.

   ANS: B   DIF: I   OBJ: 2-1.3

9. A 5 kg object has less inertia than a ____ object. (1 kg = 1,000 g)
   a. 4 kg
   b. 6,000 g
   c. 2 kg
   d. 1,500 g

   ANS: B   DIF: I   OBJ: 2-2.2

10. According to Newton's first law of motion, a moving object that is not acted on by an unbalanced force will
   a. remain in motion.
   b. eventually come to a stop.
   c. transfer its energy to another object.
   d. accelerate in the absence of friction.

   ANS: A   DIF: I   OBJ: 2-2.1

11. An astronaut uses a jet of nitrogen to maneuver in space. As the nitrogen is expelled, it
   a. exerts a reaction force on its container.
   b. causes the astronaut to accelerate.
   c. illustrates Newton's third law.
   d. All of the above

   ANS: D   DIF: I   OBJ: 2-2.3

12. Imagine that you are holding a 6 N book motionless in your hand. Which of the following is true?
   a. The book has a mass of 6 N.
   b. The total downward force on the book is 12 N.
   c. Your hand exerts an upward force of 6 N on the book.
   d. This situation illustrates unbalanced forces.

   ANS: C   DIF: I   OBJ: 2-2.1

13. Galileo proved that the rate at which an object falls
   a. increases with an increase in mass.
   b. decreases with an increase in mass.
   c. is not affected by the mass.
   d. increases with a decrease in mass.

   ANS: C   DIF: I   OBJ: 2-1.1

14. All objects accelerate toward the Earth at a rate of 9.8 m/s/s. This means that for every second that an object falls, its downward velocity
   a. increases by 9.8 m/s.
   b. decreases by 9.8 m/s.
   c. stays at 9.8 m/s.
   d. None of the above

   ANS: A   DIF: I   OBJ: 2-1.1

15. A ball is dropped from a rooftop. What is the ball's velocity after 3 s? (Assume that there is no air resistance.)
    a. 0 m/s
    b. 9.8 m/s
    c. 19.6 m/s
    d. 29.4 m/s

    ANS: D   DIF: II   OBJ: 2-1.1

16. Air resistance is
    a. sliding friction.
    b. rolling friction.
    c. fluid friction.
    d. static friction.

    ANS: C   DIF: I   OBJ: 2-1.1

17. The amount of air resistance acting on an object depends on the object's
    a. size and shape.
    b. mass and weight.
    c. density and mass.
    d. None of the above

    ANS: A   DIF: I   OBJ: 2-1.1

18. An apple falls from a tree. The gravitational force on the apple is 1 N. If air resistance is 0.1 N, what is the net force on the apple?
    a. 0.1 N
    b. 0.9 N
    c. 1.0 N
    d. 1.1 N

    ANS: B   DIF: II   OBJ: 2-1.1

19. As long as the net force on a falling object is NOT zero, the object
    a. falls at a constant velocity.
    b. accelerates downward.
    c. is pushed back up.
    d. does not move.

    ANS: B   DIF: I   OBJ: 2-1.1

20. Terminal velocity is reached when the net force on a falling object reaches
    a. −1 N.
    b. 0 N.
    c. 1 N.
    d. 9.8 N.

    ANS: B   DIF: I   OBJ: 2-1.1

21. Parachutes slow skydivers to a safer terminal velocity because parachutes
    a. increase air resistance.
    b. decrease air resistance.
    c. increase gravitational pull.
    d. decrease gravitational pull.

    ANS: A   DIF: I   OBJ: 2-1.1

22. Astronauts float inside the space shuttle because they
    a. are massless.
    b. have no gravitational force acting on them.
    c. are in free fall.
    d. are weightless.

    ANS: C   DIF: I   OBJ: 2-1.2

23. An orbit is formed when the shuttle
    a. moves forward.
    b. is in free fall.
    c. is pulled down by gravity.
    d. All of the above

    ANS: D   DIF: I   OBJ: 2-1.3

24. The shuttle traveling in space around the Earth is an example of
    a. orbiting.
    b. centripetal force.
    c. projectile motion.
    d. All of the above

    ANS: D   DIF: I   OBJ: 2-1.3

25. An example of an object in projectile motion is
    a. a leaping frog.
    b. a game of billiards.
    c. riding a bicycle.
    d. pushing a shopping cart.

    ANS: A   DIF: I   OBJ: 2-1.4

26. ____ is a measure of inertia.
    a. Distance
    b. Mass
    c. Speed
    d. Velocity

    ANS: B   DIF: I   OBJ: 2-2.1

27. Newton's second law of motion states that an object's acceleration
    a. increases as its mass decreases and as the force acting on it increases.
    b. decreases as its mass decreases and as the force acting on it increases.
    c. increases as its mass increases and as the force acting on it increases.
    d. decreases as its mass increases and as the force acting on it increases.

    ANS: A   DIF: I   OBJ: 2-2.1

28. The acceleration of an object
    a. is not related to the direction that the force was applied.
    b. is always in the direction opposite to the direction that the force was applied.
    c. is always in the same direction as the force.
    d. None of the above

    ANS: C   DIF: I   OBJ: 2-2.1

29. Use Newton's second law of motion to calculate the acceleration of a 7 kg mass if a force of 68.6 N acts on it?
    a. 0.1 m/s/s
    b. 9.8 m/s/s
    c. 68.6 m/s/s
    d. 480.2 m/s/s

    ANS: B   DIF: II   OBJ: 2-2.1

30. What force is necessary to accelerate a 1,250 kg car at a rate of 40 m/s/s?
    a. 31.25 N
    b. 40.0 N
    c. 1,250 N
    d. 50,000 N

    ANS: D   DIF: II   OBJ: 2-2.1

31. Use Newton's second law of motion to calculate the mass of an object when a force of 34 N accelerates the object 4 m/s/s?
    a. 0.12 kg
    b. 8.5 kg
    c. 38.0 kg
    d. 136 kg

    ANS: B   DIF: II   OBJ: 2-2.1

32. How much force is needed to accelerate a 70 kg rider and her 200 kg motor scooter at 4 m/s/s?
    a. 270 N
    b. 280 N
    c. 800 N
    d. 1,080 N

    ANS: D   DIF: II   OBJ: 2-2.1

33. Newton's third law of motion states that whenever one object exerts a force on a second object,
    a. the second object exerts an equal and opposite force on a third object.
    b. the first object is unaffected by that force.
    c. the second object exerts an equal and opposite force on the first object.
    d. the second object exerts a less powerful force on the first object.

    ANS: C   DIF: I   OBJ: 2-2.1

34. Newton's third law of motion states that if a force is exerted on an object, another force occurs that
    a. is equal in size and opposite in direction.
    b. is in the same direction and size.
    c. is equal in speed and opposite in direction.
    d. is in the same direction and speed.

    ANS: A   DIF: I   OBJ: 2-2.1

35. When a swimmer swims through water,
    a. the action force could be the swimmer's hands and feet pushing on the water.
    b. the reaction force could be the water pushing on the hands and feet.
    c. the reaction force is what moves the swimmer forward.
    d. All of the above

    ANS: D   DIF: I   OBJ: 2-2.1

36. Action and reaction force pairs occur
    a. only when there is motion.
    b. only when there is no motion.
    c. whether there is motion or not.
    d. only when the forces are unbalanced.

    ANS: C   DIF: I   OBJ: 2-2.1

37. You sitting in a chair is an example of Newton's
    a. second law.
    b. first and third laws.
    c. second and third laws.
    d. first and fourth laws.

    ANS: B   DIF: I   OBJ: 2-2.1

38. _____ do NOT act on the same object.
    a. Force pairs
    b. Gravitational forces
    c. Centripetal forces
    d. Inertial forces

    ANS: A   DIF: I   OBJ: 2-2.1

39. The net force of unbalanced force pairs is _____.
    a. positive.
    b. negative.
    c. either positive or negative.
    d. zero.

    ANS: C   DIF: I   OBJ: 2-2.1

40. Which of the following is NOT an example of Newton's third law of motion?
    a. hitting a baseball with a bat
    b. sitting in a chair
    c. an apple falling from a tree
    d. none of the above

    ANS: D   DIF: I   OBJ: 2-2.1

You have four vehicles, all driving at the same velocity side-by-side on a four-lane highway. They are a fully-loaded truck, an empty truck, a midsize van, and a small car.

41. Which one has the MOST momentum?
    a. the fully-loaded truck
    b. the empty truck
    c. the midsize van
    d. the small car

    ANS: A   DIF: I   OBJ: 2-2.2

42. If all the vehicles brake at the same time because there is a collision ahead, which one will come to a complete stop FIRST?
    a. the fully-loaded truck
    b. the empty truck
    c. the midsize van
    d. the small car

    ANS: D   DIF: I   OBJ: 2-2.2

43. The momentum before a collision is
    a. less than the momentum after the collision.
    b. equal to the momentum after the collision.
    c. more than the momentum after the collision
    d. completely lost after the collision.

    ANS: B   DIF: I   OBJ: 2-2.3

44. Which of the following games uses conservation of momentum?
    a. billiards
    b. bowling
    c. baseball
    d. all of the above

    ANS: D   DIF: I   OBJ: 2-2.3

45. Conservation of momentum is explained by Newton's
    a. first law of motion.
    b. second law of motion.
    c. third law of motion.
    d. fourth law of motion.

    ANS: C   DIF: I   OBJ: 2-2.3

46. Inertia is used when explaining Newton's
    a. first law of motion.
    b. second law of motion.
    c. third law of motion.
    d. law of conservation of momentum.

    ANS: A   DIF: I   OBJ: 2-2.1

47. In billiards,
    a. an action force makes a billiard ball move.
    b. an action force makes a cue ball move.
    c. a reaction force stops the cue ball when it comes in contact with a billiard ball.
    d. All of the above

    ANS: D   DIF: I   OBJ: 2-2.3

48. Jumping beans jump when a small insect larva inside the bean suddenly moves, hitting the inside
    of the shell. This is an exchange of
    a. velocity.
    b. inertia.
    c. momentum.
    d. acceleration.

    ANS: C   DIF: I   OBJ: 2-2.3

49. After a motor pulls the roller coaster cars up the first hill, what keeps the cars moving up and
    over the following hills, turns, and loops?
    a. gravity
    b. inertia
    c. acceleration
    d. all of the above

    ANS: D   DIF: I   OBJ: 2-1.1

50. Suppose you are playing a very unusual game of billiards. All the balls are different sizes. If the
    cue ball were to collide with a billiard ball twice its size, what would happen?
    a. The cue ball would stop and the billiard ball would not move.
    b. The cue ball would roll backward and the billiard ball would not move.
    c. The cue ball would roll backward and the billiard ball would move away from the cue
       ball slower than the cue ball's initial speed.
    d. The billiard ball would move away at twice the speed.

    ANS: C   DIF: III   OBJ: 2-2.3

51. Catapults create
    a. free fall.
    b. orbiting.
    c. projectile motion.
    d. gravitational force.

    ANS: C   DIF: I   OBJ: 2-1.4

52. Terminal velocity
    a. is a constant velocity.
    b. has a balanced net force.
    c. is a result of air resistance.
    d. All of the above

    ANS: D   DIF: I   OBJ: 2-1.1

53. Which of the following will have the greatest air resistance?
   a. an acorn
   b. a crumpled-up sheet of paper
   c. an 8 1/2" × 11" sheet of paper
   d. an apple

   ANS: C   DIF: I   OBJ: 2-1.1

54. Because of projectile motion, when aiming at a target you should always aim
   a. above the bull's eye.
   b. below the bull's eye.
   c. to the left of the bull's eye.
   d. to the right of the bull's eye.

   ANS: A   DIF: I   OBJ: 2-1.4

55. The downward acceleration of a thrown object in projectile motion is
   a. less than the acceleration of a vertically falling object because of the horizontal part of
      the projectile motion.
   b. greater than the acceleration of a vertically falling object because of the horizontal part
      of the projectile motion.
   c. identical to that of a vertically falling object regardless of the horizontal part of the
      projectile motion.
   d. practically nonexistent because of the horizontal part of the projectile motion.

   ANS: C   DIF: II   OBJ: 2-1.4

56. Which of the following is the unbalanced force that maintains circular motion for an object in
   orbit?
   a. projectile motion
   b. centripetal force
   c. free fall
   d. all of the above

   ANS: B   DIF: I   OBJ: 2-1.3

57. Is it just as hard to catch a thrown bowling ball as it is to throw it?
   a. The bowling ball has more inertia while in motion so it's harder to catch it.
   b. The bowling ball has the same inertia whether it's standing still or moving, so throwing it
      and catching it are both equally difficult.
   c. The bowling ball has less inertia while in motion, so it's easier to catch than it is to
      throw.
   d. Inertia has nothing to do with how easy or hard it is to throw or catch a bowling ball. It
      just depends on how strong you are.

   ANS: B   DIF: I   OBJ: 2-2.1

58. If a car driver suddenly makes a sharp turn, the passenger slides to the side of the car because of
   a. inertia.
   b. free fall.
   c. gravity.
   d. friction.

   ANS: A   DIF: I   OBJ: 2-2.1

59. When an airplane takes off, you tend to fall backward because of
   a. air resistance.
   b. inertia.
   c. gravity.
   d. friction.

   ANS: B   DIF: I   OBJ: 2-2.1

60. When you bump into someone standing still, you can knock them over because of
    a. projectile motion.
    b. momentum.
    c. gravity.
    d. friction.

    ANS: B   DIF: I   OBJ: 2-2.3

61. When two bumper cars collide, the force exerted on each car causes a change in the momentum for each car. The total _____ for both cars is the same before and after the collision.
    a. terminal velocity
    b. inertia
    c. gravity
    d. momentum

    ANS: D   DIF: I   OBJ: 2-2.3

62. If you pull your hands back as you catch a fast ball, it tends to hurt less than if you keep your hands still. The momentum of the ball that is transferred to your hand is reduced because
    a. the velocity of your hand reduces the impact from the velocity of the ball because they both move in the same direction.
    b. the ball encounters more air resistance.
    c. the inertia of the ball decreases.
    d. the ball maintains projectile motion for a little longer, decreasing its velocity.

    ANS: A   DIF: I   OBJ: 2-2.3

63. In the Olympic sport of discus throwing, a discus thrower will spin around in a circle with the discus and suddenly stop spinning and let go of the discus. Instead of stopping with the thrower, the discus flies off into the air because of its
    a. gravitational force.
    b. inertia.
    c. mass.
    d. centripetal force.

    ANS: B   DIF: I   OBJ: 2-2.3

64. The curved path traveled by a thrown baseball is known as
    a. orbiting.
    b. centripetal acceleration.
    c. projectile motion.
    d. centripetal force.

    ANS: C   DIF: I   OBJ: 2-1.4

65. Using Newton's second law of motion, what is a skater's acceleration when a 50 kg skater pushes off from a wall with a force of 200 N?
    a. $\frac{1}{4}$ m/s/s
    b. 4 m/s/s
    c. 50 m/s/s
    d. 10,000 m/s/s

    ANS: B   DIF: I   OBJ: 2-2.1

66. Which of the following has the most momentum?
    a. an ant moving at 1 m/s
    b. a bird flying at 4 m/s
    c. a cat moving at 13 m/s
    d. you riding your bike at 12 m/s

    ANS: D   DIF: I   OBJ: 2-2.3

67. Orbiting objects appear to be weightless because they are
    a. in free fall.   c. outside of Earth's atmosphere.
    b. weightless.   d. in space.

    ANS: A   DIF: I   OBJ: 2-1.2

68. A satellite orbits Earth because
    a. it is caught in Earth's gravitational pull, like a tractor beam.
    b. it does not have enough fuel to start moving.
    c. it is moving forward while it is in free fall toward Earth.
    d. it is at rest.

    ANS: C   DIF: I   OBJ: 2-1.3

69. Which of the following is an example of free fall?
    a. a skydiver falling from an airplane
    b. a floating astronaut in orbit around Earth
    c. a ball falling from a rooftop
    d. tossing a set of keys to a friend

    ANS: B   DIF: I   OBJ: 2-1.2

70. Although astronauts appear to be weightless, they are not because
    a. mass increases as distance increases.
    b. they can only be weightless outside the solar system.
    c. they are in orbit. If they were to leave orbit, they would be weightless.
    d. their mass does not change, so the astronauts would always have some gravitational
        attraction to objects—and therefore, they have weight.

    ANS: D   DIF: I   OBJ: 2-1.2

71. The KC-135 Vomit Comet is able to simulate an astronaut's apparent weightlessness in space by
    diving toward the ground at a 45° angle because
    a. it combines the motions of free fall with moving forward.
    b. gravitational force changes with angle.
    c. gravitational force depends on speed.
    d. gravitational force is zero when the distance changes quickly enough.

    ANS: A   DIF: I   OBJ: 2-1.2

72. Planets stay in orbit around the sun because
    a. unbalanced forces act on them.
    b. two motions combine to cause orbiting.
    c. the sun's gravitational force provides a centripetal force on the planets.
    d. All of the above

    ANS: D   DIF: I   OBJ: 2-1.3

73. Centripetal force on an object acts in the
    a. same direction as gravitational force.
    b. direction opposite the direction of gravitational force.
    c. same direction as the path of the object.
    d. direction opposite the object's path.

    ANS: A   DIF: II   OBJ: 2-1.3

74. Any object in circular motion
    a. experiences free fall toward Earth.   c. is weightless while in orbit.
    b. constantly changes direction.   d. is massless while in orbit.

    ANS: B   DIF: I   OBJ: 2-1.3

75. Projectile motion
    a. has one component—horizontal.
    b. has one component—vertical.
    c. has two components—horizontal and vertical.
    d. cannot be broken down into directional components.

    ANS: C   DIF: I   OBJ: 2-1.4

76. The components of projectile motion
    a. depend greatly on each other.
    b. have no effect on each other.
    c. have a little effect on each other, but it is negligible.
    d. cannot be combined.

    ANS: B   DIF: I   OBJ: 2-1.4

77. Quarterbacks must aim higher than their target because the ball's
    a. vertical velocity increases because gravity causes it to accelerate downward.
    b. horizontal velocity increases because it is accelerating.
    c. vertical velocity is constant.
    d. horizontal velocity is constant.

    ANS: A   DIF: I   OBJ: 2-1.4

COMPLETION

1. An object in motion tends to stay in motion because it has _____.
   (inertia or terminal velocity)

   ANS: inertia   DIF: I   OBJ: 2-2.1

2. Falling objects stop accelerating at _____. (free fall or terminal velocity)

   ANS: terminal velocity   DIF: I   OBJ: 2-1.1

3. _____ is the path that a thrown object follows. (Free fall or Projectile motion)

    ANS: Projectile motion   DIF: I   OBJ: 2-1.4

4. A property of moving objects that depends on mass and velocity is _____.
    (inertia or momentum)

    ANS: momentum   DIF: I   OBJ: 2-2.1

5. _____ only occurs when there is no air resistance. (Momentum or Free fall)

    ANS: Free fall   DIF: I   OBJ: 2-1.1

6. When air resistance exactly matches the downward force of gravity, a falling object stops
    accelerating and reaches _____. (free fall or terminal velocity)

    ANS: terminal velocity   DIF: I   OBJ: 2-1.1

7. An object at rest tends to remain at rest. This property is called _____.
    (momentum or inertia)

    ANS: inertia   DIF: I   OBJ: 2-2.1

8. _____ can occur only in a vacuum or in space. (Projectile motion or Free fall)

    ANS: Free fall   DIF: I   OBJ: 2-1.1

9. A car accelerates from 60 km/h to a constant speed of 80 km/h. The car has increased its
    _____. (momentum or terminal velocity)

    ANS: momentum   DIF: I   OBJ: 2-2.2

10. The horizontal movement given to an arrow by a bow is one component of
    _____. (projectile motion or free fall)

    ANS: projectile motion   DIF: I   OBJ: 2-1.4

11. Because air particles have mass, they also have inertia that resists movement, which is called
    _____.

    ANS: air resistance   DIF: I   OBJ: 2-1.1

12. Traveling in a circular or a nearly circular path around another object is called
    _____.

    ANS: orbiting   DIF: I   OBJ: 2-1.3

13. The unbalanced force that causes objects to move in a circular path is called a

 _____.

    ANS: centripetal force   DIF: I   OBJ: 2-1.3

14. _____ motion is motion that is parallel to the ground.

    ANS: Horizontal   DIF: I   OBJ: 2-1.4

15. _____ motion is motion that is perpendicular to the ground.

    ANS: Vertical   DIF: I   OBJ: 2-1.4

16. The law of _____ of momentum states that when two or more objects interact, they may exchange momentum, but the total amount of momentum stays the same.

    ANS: conservation   DIF: I   OBJ: 2-2.3

17. If you're riding a horse and your horse suddenly stops just before a hurdle, you will most likely continue moving forward. This is explained in Newton's _____ law of motion.

    ANS: first   DIF: I   OBJ: 2-2.1

18. _____ causes acceleration to stop at terminal velocity.

    ANS: Air resistance   DIF: I   OBJ: 2-1.1

19. _____ provides the centripetal force that keeps objects in orbit.

    ANS: Gravitational force   DIF: I   OBJ: 2-1.3

SHORT ANSWER

1. How does air resistance affect the acceleration of falling objects?

   ANS:
   Air resistance slows or stops the acceleration of falling objects.

   DIF: I   OBJ: 2-1.3

2. Explain why an astronaut in an orbiting spaceship floats.

   ANS:
   An astronaut on an orbiting spaceship floats because both the astronaut and the spaceship are in free fall. Since both fall at the same rate, the astronaut floats inside. The astronaut has no sensation of falling.

   DIF: II   OBJ: 2-1.2

3. How is an orbit formed?

   ANS:
   An orbit is formed by combining two motions: a forward motion and free fall toward Earth. The path that results is a curve that matches the curve of Earth's surface.

   DIF: II   OBJ: 2-1.3

4. Think about a sport you play that involves a ball. Identify at least four different instances in which an object is in projectile motion.

   ANS:
   Accept all reasonable answers. A basketball example: a player jumping to dunk the ball; a ball passed from one player to another; a ball shot toward the basket; a ball bounced on the floor.

   DIF: I   OBJ: 2-1.4

5. How is inertia related to Newton's first law of motion?

   ANS:
   Newton's first law says that matter resists any change in motion. Inertia is the *tendency* of objects (matter) to resist changes in motion. Newton's first law is also known as the law of inertia.

   DIF: I   OBJ: 2-2.1

6. Name two ways to increase the acceleration of an object.

   ANS:
   You can increase the acceleration of an object by increasing the force causing the acceleration or by reducing the object's mass.

   DIF: I   OBJ: 2-2.1

7. If the acceleration due to gravity were somehow doubled to 19.6 m/s/s, what would happen to your weight?

   ANS:
   If the acceleration due to gravity were doubled, your weight would double. This is because of Newton's second law: $F = ma$. Weight is the force due to the acceleration on mass. If acceleration is doubled and mass remains the same, the force (weight) is doubled, too.

   DIF: I   OBJ: 2-2.1

8. Name three action and reaction force pairs involved in doing your homework. Name what object is exerting and what object is receiving the forces.

ANS:
Accept all reasonable answers. Sample answers include using a pencil or pen (action: hand pushing on pencil; reaction: pencil pushing back on hand OR action: pencil pushing on paper; reaction: paper pushing on pencil).

DIF: I   OBJ: 2-2.1

9. Which has more momentum, a mouse running at 1 m/s north or an elephant walking at 3 m/s east? Explain your answer.

ANS:
The elephant—it has both a greater mass and greater velocity.

DIF: I   OBJ: 2-2.2

10. When a truck pulls a trailer, the trailer and truck accelerate forward even though the action and reaction forces are the same size but in opposite directions. Why don't these forces balance each other out?

ANS:
The action and reaction forces do not balance each other because the forces are acting on two different objects. Because they act on two different objects, you cannot combine them to determine a net force.

DIF: II   OBJ: 2-2.2

11. Explain why a ball moves in a straight line as it rolls across a table but follows a curved path once it rolls off the edge of the table.

ANS:
A ball rolling across the table has only horizontal motion. Once the ball rolls off the edge, gravity pulls it downward, giving it both vertical and horizontal motion, which causes a curved path.

DIF: I   OBJ: 2-1.4

12. Explain why results differ on the moon and on Earth when a hammer and a feather are dropped from the same height at exactly the same time.

ANS:
On the moon, both will hit the ground at the same time because there is no atmosphere and no air resistance. On Earth, the hammer will hit the ground first because the feather will be slowed much more by air resistance.

DIF: I   OBJ: 2-1.1

13. How does Newton's second law explain why it is easier to push a bicycle than to push a car with the same acceleration?

ANS:
The bicycle has a smaller mass, so a smaller force is required to give it the same acceleration as the car.

DIF: I   OBJ: 2-2.1

14. What are two ways that you can increase the acceleration of a loaded grocery cart?

ANS:
You can increase the force applied to the cart, or you can decrease the mass of the cart by removing some of the objects from it.

DIF: I   OBJ: 2-2.1

15. How does Newton's third law explain how a rocket takes off?

ANS:
The hot gases expelled from the back of the rocket produce a reaction force on the rocket that lifts and accelerates the rocket.

DIF: I   OBJ: 2-2.1

16. Explain how an orbit is formed.

ANS:
An Earth orbit is formed by combining the forward motion of the orbiting object with free fall toward Earth. The path that results is a curve that follows the curve of the Earth.

DIF: I   OBJ: 2-1.3

17. Describe how gravity and air resistance combine when an object reaches terminal velocity.

ANS:
Gravity and air resistance combine to give a net force of zero on a falling object. When this happens, the object stops accelerating downward and has reached its terminal velocity.

DIF: I   OBJ: 2-1.1

18. Explain why friction can make observing Newton's first law of motion difficult.

ANS:
Friction is a force that opposes the motion of objects. Friction is what slows the motion of moving objects so you don't see objects moving forever in a straight line.

DIF: I   OBJ: 2-2.1

19. Use the following terms to create a concept map: *gravity, free fall, terminal velocity, projectile motion, air resistance.*

ANS:

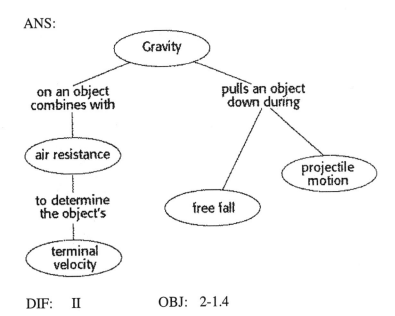

DIF: II   OBJ: 2-1.4

20. During a shuttle launch, about 830,000 kg of fuel is burned in 8 minutes. The fuel provides the shuttle with a constant thrust, or push off the ground. How does Newton's second law of motion explain why the shuttle's acceleration increases during takeoff?

ANS:
Newton's second law: $a = F \div m$. During takeoff, the shuttle burns fuel and therefore loses mass. However, the upward force on the shuttle remains the same. So the shuttle's acceleration increases because its mass constantly gets smaller during takeoff.

DIF: II   OBJ: 2-2.1

21. When using a hammer to drive a nail into wood, you have to swing the hammer through the air with a certain velocity. Because the hammer has both mass and velocity, it has momentum. Describe what happens to the hammer's momentum after the hammer hits the nail.

ANS:
When the hammer hits the nail, the hammer stops. Its momentum is transferred to the nail, driving it into the wood. Momentum is also transferred from the nail to the wood and to the work bench or table top.

DIF: II   OBJ: 2-2.3

22. Suppose you are standing on a skateboard or on in-line skates and you toss a backpack full of heavy books toward your friend. What do you think will happen to you and why? Explain your answer in terms of Newton's third law of motion.

ANS:
You will move away from your friend (in the direction opposite from where you throw the backpack). The action force is you pushing the backpack toward your friend. The reaction force is the backpack pushing you away from your friend.

DIF: II   OBJ: 2-2.1

23. A 12 kg rock falls from rest off a cliff and hits the ground in 1.5 s.
    a. Ignoring air resistance, what is the rock's velocity just before it hits the ground?
    b. What is the rock's weight after it hits the ground? (Hint: Weight is a measure of the gravitational force on an object.)

ANS:
    a. $\Delta v = g \times t = 9.8$ m/s/s $\times 1.5$ s $= 14.7$ m/s
    b. $F = m \times a = 12$ kg $\times 9.8$ m/s/s $= 117.6$ N

DIF: II   OBJ: 2-1.1

24. The picture below shows a common desk toy. If you pull one ball up and release it, it hits the balls at the bottom and comes to a stop. In the same instant, the ball on the other side swings up and repeats the cycle. How does conservation of momentum explain how this toy works?

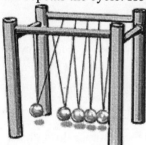

ANS:
The law of conservation of momentum: When two or more objects interact, the total amount of momentum must stay the same. The ball moving in the air has a certain amount of momentum, and the balls at rest have no momentum. When the moving ball hits the balls at rest, all of its momentum is transferred to them, and it comes to a stop. The momentum is transferred from ball to ball until it reaches the ball on the other end. The ball on the other end keeps all the momentum, and it moves away from the other balls.

DIF: II   OBJ: 2-2.3

25. You are a passenger in a car that is moving rapidly down a straight road. As the driver makes a sharp left turn, you are pressed against the right side of the car. Explain why this happens.

ANS:
The car changes direction, but because of inertia your body tends to continue traveling forward. Thus, you are pressed against the right side of the car.

DIF: I   OBJ: 2-2.1

26. Use Newton's third law to explain how a person using a hammer to drive a nail into a board is demonstrating conservation of momentum.

ANS:
Conservation of momentum can be explained by Newton's third law: Whenever one object exerts a force on a second object, the second object exerts an equal and opposite force on the first. When the hammer strikes the nail, the hammer stops moving. The hammer's momentum is transferred to the nail, which is driven into the board. Thus, the objects exchange momentum, and total momentum is conserved.

DIF: II   OBJ: 2-2.3

27. Why, in a classroom, would a feather dropped from the same height as an acorn fall to Earth more slowly than the acorn?

ANS:
Both would fall at the same rate in a vacuum. However, there is no vacuum in the classroom, so the feather's larger surface area causes it to experience greater upward force due to air resistance than the acorn does, and the feather falls more slowly.

DIF: I   OBJ: 2-1.1

28.
   a. Imagine that you and a friend are standing on the roof of a tall building. You throw a snowball straight out from the building. The snowball moves away from the building at a velocity of 25 m/s. What type of motion does the snowball exhibit? Explain.
   b. At exactly the same time that you throw your snowball, your friend drops a rubber ball from the roof. It falls with an acceleration of 9.8 m/s/s. Assuming the snowball and the rubber ball are opposed by equal amounts of air resistance, which would hit the ground first—the snowball or the rubber ball? Explain your answer.

ANS:

   a. The snowball exhibits projectile motion because it follows a curved path when it is thrown from the building.
   b. The snowball is a projectile, so its horizontal and vertical motion components are independent. As the snowball moves away from the building at a velocity of 25 m/s, it also accelerates toward the Earth at a rate of 9.8 m/s/s. The rubber ball accelerates toward the Earth at the same rate. Both objects would hit the ground at the same time.

DIF: II   OBJ: 2-1.4

29. Use the following terms to complete the concept map below: *direction, air resistance, weight, terminal velocity, mass.*

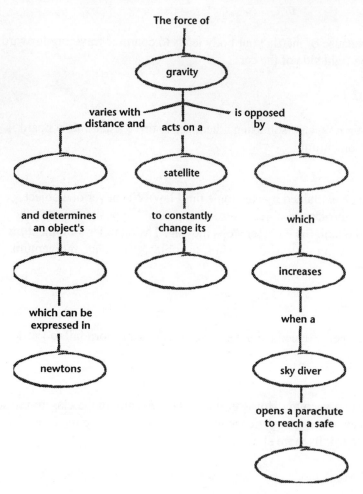

The force of

gravity

varies with distance and   acts on a   is opposed by

satellite

and determines an object's   to constantly change its   which

increases

which can be expressed in   when a

newtons   sky diver

opens a parachute to reach a safe

ANS:

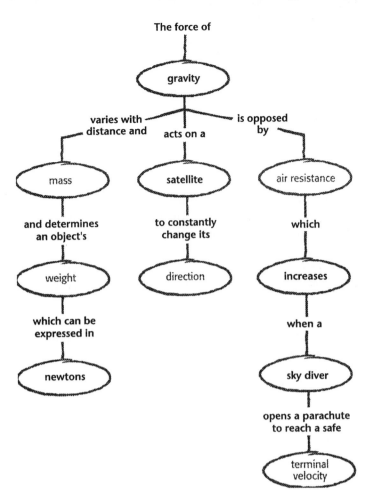

The force of

gravity

varies with distance and   acts on a   is opposed by

mass   satellite   air resistance

and determines an object's   to constantly change its   which

weight   direction   increases

which can be expressed in   when a

newtons   sky diver

opens a parachute to reach a safe

terminal velocity

DIF: II   OBJ: 2-1.1

30.   A rock climber dislodges a stone while climbing a mountain. The stone falls straight down, taking exactly 3.5 s to hit the ground. Ignoring air resistance, how fast was the stone traveling when it hit the ground? Show your work.

ANS:
$v = g \times t = 9.8 \text{ m/s/s} \times 3.5 \text{ s} = 34.3 \text{ m/s}$
At the time of impact, the stone was traveling at a velocity of 34.3 m/s.

DIF: II   OBJ: 2-1.1

ESSAY

1. A basketball player throws a basketball toward a basketball net. The basketball hits the backboard, bounces back, comes down, and hits the rim of the basket. Then it bounces off the rim and falls to the ground where it bounces some more before it comes to a stop. Describe the forces that were present while applying Newton's first and third laws of motion to this situation.

ANS:
Accept all reasonable answers. Sample answer: *Newton's first law of motion* states that an object at rest remains at rest unless acted on by an unbalanced force. The basketball player's hand is the unbalanced force that starts the ball in motion. This law also states that an object in motion remains in motion at constant speed and in a straight line unless acted upon by an unbalanced force. The backboard, the rim, and the ground are all unbalanced forces that affect the constant speed and straight line of the ball. Gravity and the fluid friction of air also affect the speed and straight line of the ball causing it to slow down and fall toward the ground.

The *third law of motion* states that whenever one object exerts a force on a second object, the second object exerts an equal and opposite force on the first. When the ball hits the backboard, the backboard exerts a force back onto the ball causing it to bounce back, gravity causes it to fall to the rim. The rim exerts a force causing the ball to bounce back, and gravity pulls it to the ground. The ground exerts a force causing the ball to bounce until gravity and friction overcome the initial force exerted by the basketball player's hand and the speed of the ball. The ball will stop and remain at rest until another unbalanced force puts it in motion again.

DIF: II   OBJ: 2-2.1

MULTIPLE CHOICE

1. The design of a wing
   a. causes the air above the wing to travel faster than the air below the wing.
   b. helps create lift.
   c. creates a low-pressure zone above the wing.
   d. All of the above

   ANS: D   DIF: I   OBJ: 3-3.2

2. An object displaces a volume of fluid that
   a. is equal to its own volume.   c. is greater than its own volume.
   b. is less than its own volume.   d. is more dense than itself.

   ANS: A   DIF: I   OBJ: 3-2.2

3. Fluid pressure is always directed
   a. up.   c. sideways.
   b. down.   d. in all directions.

   ANS: D   DIF: I   OBJ: 3-1.1

4. If an object weighing 50 N displaces a volume of water with a weight of 10 N, what is the buoyant force on the object?
   a. 60 N   c. 40 N
   b. 50 N   d. 10 N

   ANS: D   DIF: II   OBJ: 3-2.1

5. A helium-filled balloon will float in air because
   a. there is more air than helium.   c. helium is as dense as air.
   b. helium is less dense than air.   d. helium is more dense than air.

   ANS: B   DIF: I   OBJ: 3-2.3

6. Materials that can flow to fit their containers include
   a. gases.   c. both gases and liquids.
   b. liquids.   d. neither gases nor liquids.

   ANS: C   DIF: I   OBJ: 3-1.1

7. Of the following, where would atmospheric pressure be greatest?
   a. at sea level   c. at the altitude at which planes fly
   b. on a mountaintop   d. at the outer edge of the atmosphere

   ANS: A   DIF: I   OBJ: 3-1.2

8. If a fluid flows from area A to area B, then area A must be an area of greater
   a. temperature.
   c. volume.
   b. mass.
   d. pressure.

   ANS: D   DIF: I   OBJ: 3-1.3

9. Which statement best explains why air bubbles in water rise to the surface?
   a. Liquids cannot be compressed very much.
   b. Water is about 1,000 times denser than air.
   c. Pressure is the amount of force exerted on a given area.
   d. The weight of the atmosphere pushes down on the water.

   ANS: B   DIF: I   OBJ: 3-2.3

10. Fluid pressure is exerted evenly in all directions, which explains why
    a. some objects float.
    c. bubbles are round.
    b. birds and airplanes can fly.
    d. objects seem to weigh less in water.

    ANS: C   DIF: I   OBJ: 3-1.1

11. Which of the following does NOT affect the amount of lift on an airplane's wing?
    a. turbulence
    c. wing surface area
    b. gravity
    d. the airplane's speed

    ANS: B   DIF: I   OBJ: 3-3.2

12. Which of the following would NOT affect the level at which a cargo ship floats in a body of
    water?
    a. the depth of the water
    c. the mass of the ship's cargo
    b. the shape of the ship
    d. the density of the ship's material

    ANS: A   DIF: I   OBJ: 3-2.3

13. The SI unit for pressure is
    a. Pa.
    c. N.
    b. Pr.
    d. K.

    ANS: A   DIF: I   OBJ: 3-1.1

14. Which of the following is NOT a fluid?
    a. water
    c. oil
    b. ice
    d. oxygen

    ANS: B   DIF: I   OBJ: 3-1.1

15. What is the pressure exerted on the floor by a 3,000 N crate with an area of 2 m^2?
    a. 1,500 Pa
    c. 3,000 Pa
    b. 3,000 Pr
    d. 6,000 N

    ANS: A   DIF: II   OBJ: 3-1.1

16. What is the weight of a rock with an area of 10 m² that exerts a pressure of 250 Pa in correct SI units?
    a. 25 m²Pa
    c. 2,500 m²Pa
    b. 250 N
    d. 2,500 N

    ANS: D   DIF: II   OBJ: 3-1.1

17. Gases exert pressure
    a. randomly in varying directions.
    c. horizontally.
    b. evenly in all directions.
    d. vertically.

    ANS: B   DIF: I   OBJ: 3-1.1

18. At sea level, the atmosphere exerts a pressure of approximately
    a. 101,000 Pa.
    c. 10 N on every square centimeter.
    b. 101,000 N on every square meter.
    d. All of the above

    ANS: D   DIF: I   OBJ: 3-1.1

19. You heat an empty aluminum soda can for several minutes with a high-wattage hair dryer, causing the heated, "energized" particles of air inside the can to escape through the opening. Quickly, you seal the can's opening with strong tape. As the can cools, atmospheric pressure will
    a. cause the can to explode.
    c. crush the can.
    b. do nothing to the can.
    d. allow the can to float.

    ANS: C   DIF: I   OBJ: 3-1.1

20. At the top of Mount Everest, atmospheric pressure is about
    a. a quarter of what it is at sea level.
    c. half of what it is at sea level.
    b. a third of what it is at sea level.
    d. two-thirds of what it is at sea level.

    ANS: B   DIF: I   OBJ: 3-1.2

21. At ____ below the surface of water, pressure is about 5,000 kPa and divers below this level must wear special suits to survive the bone-crushing pressure.
    a. 10 m
    c. 500 m
    b. 100 m
    d. 5,000 m

    ANS: C   DIF: I   OBJ: 3-1.2

22. Pressure exerted on a diver 10 m below the water's surface is twice the pressure at the surface. The pressure at this depth is
    a. 50 kPa.
    c. 202 kPa.
    b. 101 kPa.
    d. 400 kPa.

    ANS: C   DIF: I   OBJ: 3-1.2

23. The *Titanic* rests 3,660 m below sea level. The water pressure exerted on the Titanic is
    a. 36 kPa.
    c. 3,660 kPa.
    b. 360 kPa.
    d. 36,600 kPa.

    ANS: D   DIF: I   OBJ: 3-1.2

24. Water is about 1,000 times more _____ than air and, therefore, exerts greater pressure than air.
    a. dense   c. buoyant
    b. fluid   d. heavy

    ANS: A   DIF: I   OBJ: 3-1.2

25. The deepest an undersea vessel (the *Trieste*) has traveled and withstood the pressure (110,000 kPa) was
    a. 10 m.   c. 1,100 m.
    b. 100 m.   d. 11,000 m.

    ANS: D   DIF: I   OBJ: 3-1.2

26. Fluids flow
    a. from regions of high pressure to low pressure.
    b. only when pressures are even.
    c. from regions of low pressure to high pressure.
    d. only under extremely warm conditions.

    ANS: A   DIF: I   OBJ: 3-1.3

27. An example of a fluid flowing during a change in pressure is
    a. sipping a drink through a straw.   c. opening a carbonated beverage.
    b. breathing.   d. All of the above

    ANS: D   DIF: I   OBJ: 3-1.3

28. Sipping a drink through a straw that has been bent and cracked is difficult because air flowing through the crack
    a. keeps the pressure high in the straw so there's not enough change in air pressure to allow the liquid to flow.
    b. causes the pressure to drop dramatically, forcing the liquid to stay down.
    c. keeps the pressure low in the straw so there's not enough change in air pressure to allow the liquid to flow.
    d. None of the above

    ANS: A   DIF: I   OBJ: 3-1.3

29. The water-pumping station in your town increases the water pressure by 20 Pa. At which of the following locations will the water pressure be increased the most?
    a. the kitchen at the pumping station
    b. a supermarket two blocks away
    c. a home 2 km away
    d. The water pressure will be the same at all of the stations.

    ANS: D   DIF: I   OBJ: 3-1.4

30. Hydraulic devices use liquids instead of gases because
    a. liquids take on the shape of their containers and gases do not.
    b. liquids can be compressed and gases cannot.
    c. gases can be compressed and liquids cannot.
    d. gases cannot exert a force but liquids can.

    ANS: C   DIF: I   OBJ: 3-1.4

31. Although the brake pads on each wheel of a car are much larger than the brake pedal the driver uses to stop a car, the pressure applied to the brake pedal is evenly exerted on all four brake pads because
    a. of hydraulics.
    b. of Pascal's principle.
    c. the brake fluid's change in pressure is equally transmitted to all four pads at the same time.
    d. All of the above

    ANS: D   DIF: I   OBJ: 3-1.4

32. An object that weighs 2 N displaces 250 mL of water, which weighs 2.5 N. What is the buoyant force on the object? Will it sink or float?
    a. 2.5 N and it will float   c. 2 N and it will float
    b. 2.5 N and it will sink   d. 2 N and it will sink

    ANS: A   DIF: II   OBJ: 3-2.2

33. An object sinks when it displaces a volume of liquid that has a weight
    a. less than the object's weight.   c. more than the object's weight.
    b. equal to the object's weight.   d. None of the above

    ANS: A   DIF: I   OBJ: 3-2.2

34. What is the density of a 20 cm^3 sample of liquid with a mass of 25 g?
    a. 0.8 g/cm^3 c. 45 g/cm^3
    b. 1.25 g/cm^3 d. 500 g/cm^3

    ANS: B   DIF: II   OBJ: 3-2.3

35. A 546 g fish displaces 420 cm^3 of water. What is the density of the fish?
    a. 0.77 g/cm^3 c. 126 g/cm^3
    b. 1.3 g/cm^3 d. 966 g/cm^3

    ANS: B   DIF: II   OBJ: 3-2.3

36. If a rock displaces 5 L of water, it means the volume of the rock is
    a. 5 cm^3.   c. 5,000 mL.
    b. 5 g.   d. 5,000 cm^3.

    ANS: D   DIF: II   OBJ: 3-2.3

37. Most substances
    a. are less dense than air.
    b. have the same density as air.
    c. are more dense than air.
    d. have less mass than an equal volume of air.

    ANS: C   DIF: I   OBJ: 3-2.3

38. Helium floats in air because
    a. it is much less dense than air.
    b. it is twice as dense as air.
    c. it contains more mass than an equal volume of air.
    d. the buoyant force is less than the weight of helium.

    ANS: A   DIF: I   OBJ: 3-2.3

39. Steel is almost eight times more dense than water. A steel ship can float in water instead of sink because of its
    a. mass.   c. shape.
    b. weight.   d. density.

    ANS: C   DIF: I   OBJ: 3-2.3

40. Because density is mass per unit volume, an increase in a steel ship's
    a. mass leads to a decrease in its density.
    b. volume leads to a decrease in its density.
    c. volume leads to an increase in its density.
    d. density is caused by an increase in its volume.

    ANS: B   DIF: I   OBJ: 3-2.3

41. A submarine can travel both on the surface of the water and underwater because it can change its
    a. interior pressure.   c. density.
    b. gravitational force.   d. volume.

    ANS: C   DIF: I   OBJ: 3-2.3

42. If you blow a steady stream of air between two sheets of paper, the papers will
    a. blow apart because the fast-moving air has a lower pressure than the air outside the pieces of paper.
    b. blow apart because the fast-moving air has a higher pressure than the air outside the pieces of paper.
    c. move toward each other because the fast-moving air has a lower pressure than the air outside the pieces of paper.
    d. move toward each other because the fast-moving air has a higher pressure than the air outside the pieces of paper.

    ANS: C   DIF: II   OBJ: 3-3.1

43. A spray gun, used to apply paint, contains two tubes inside of it: one tube comes up from the supply of paint and one supplies a flow of compressed air that blows across the top of the paint tube. This rush of air creates ____, which sucks the paint up the tube and into the stream of air.
   a. thrust
   b. drag
   c. low pressure
   d. high pressure

   ANS: C   DIF: II   OBJ: 3-3.3

44. The shape of a bird's wings can create
   a. lift.
   b. thrust.
   c. gravity.
   d. weight.

   ANS: A   DIF: I   OBJ: 3-3.2

45. Larger wings on a plane allow it to achieve greater
   a. lift.
   b. thrust.
   c. drag.
   d. speed.

   ANS: A   DIF: I   OBJ: 3-3.2

46. Jet engines can create a great deal of ____, so the wings do not have to be very big.
   a. weight
   b. thrust
   c. lift
   d. drag

   ANS: B   DIF: I   OBJ: 3-3.2

47. Flaps on airplane wings help reduce turbulence, which causes drag and can reduce
   a. weight.
   b. altitude.
   c. air resistance.
   d. lift.

   ANS: D   DIF: I   OBJ: 3-3.2

48. Bernoulli's principle can be demonstrated by using
   a. hydraulics.
   b. a Frisbee®.
   c. helium balloons.
   d. a rubber duck.

   ANS: B   DIF: I   OBJ: 3-3.3

49. In baseball, a good curveball can be explained by
   a. Pascal's principle.
   b. Archimedes' principle.
   c. Bernoulli's principle.
   d. Boyle's principle.

   ANS: C   DIF: II   OBJ: 3-3.3

50. The reason you blow round bubbles and not square ones is explained by
   a. Pascal's principle.
   b. Archimedes' principle.
   c. Bernoulli's principle.
   d. Boyle's principle.

   ANS: A   DIF: I   OBJ: 3-1.4

51. The atmosphere stretches approximately ____ above us.
    a. 10 km
    b. 15 km
    c. 50 km
    d. 150 km

    ANS: D   DIF: I   OBJ: 3-1.2

52. Eighty percent of the gases in the atmosphere are found within ____ of the Earth's surface.
    a. 1 km
    b. 5 km
    c. 10 km
    d. 50 km

    ANS: C   DIF: I   OBJ: 3-1.2

53. Pressure depends on
    a. the total amount of fluid present.
    b. the depth of the fluid.
    c. the temperature of the fluid.
    d. the compressibility of the fluid.

    ANS: B   DIF: I   OBJ: 3-1.1

54. A volume of water exerts greater pressure than the same volume of air because the water
    a. has more mass.
    b. has more weight.
    c. is more dense.
    d. All of the above

    ANS: D   DIF: I   OBJ: 3-1.1

55. Where would you notice the most pressure?
    a. 100 m above the surface of the Earth
    b. 10 m above the surface of the Earth
    c. 10 m below the surface of water
    d. 100 m below the surface of water

    ANS: D   DIF: I   OBJ: 3-1.2

56. When nurses use needles to draw a sample of blood from your arm, they begin with a needle that has the plunger pushed all the way in and then slowly pull the plunger out. This causes
    a. a decrease in pressure inside the needle and since fluids flow from high pressure to low pressure, blood flows into the needle.
    b. an increase in pressure inside the needle and since fluids flow from high pressure to low pressure, blood flows into the needle.
    c. an increase in pressure inside the needle and since fluids flow from low pressure to high pressure, blood flows into the needle.
    d. a decrease in pressure inside the needle and since fluids flow from low pressure to high pressure, blood flows into the needle.

    ANS: A   DIF: II   OBJ: 3-1.3

57. A force of 15 N is required to lift an object that is underwater. The object displaces 2 L of water. If 1 L of water weighs 10 N, what is the weight of the object out of water?
    a. 10 N
    b. 15 N
    c. 20 N
    d. 35 N

    ANS: D   DIF: II   OBJ: 3-2.2

58. A pitcher throwing a curve ball knows how to take advantage of
    a. Pascal's principle.   c. Bernoulli's principle.
    b. Archimedes' principle.   d. Boyle's principle.

    ANS: C   DIF: I   OBJ: 3-3.3

59. Earth's continental crust floats on the asthenosphere below it because the rock that makes up the Earth's continental crust is _____ the molten mantle rock.
    a. more dense than   c. the same density as
    b. less dense than   d. in a bowl-like shape, unlike

    ANS: B   DIF: I   OBJ: 3-2.3

60. Air travels _____ over the top of a wing.
    a. faster   c. the same speed as
    b. slower   d. turbulently

    ANS: A   DIF: II   OBJ: 3-3.1

61. Birds with _____ must generate more thrust in order to achieve lift.
    a. large wings   c. two wings
    b. small wings   d. triangular wings

    ANS: B   DIF: I   OBJ: 3-3.2

62. Some people who keep parakeets as pets trim the flight feathers. This makes it more difficult to fly because the size of the wing has changed, which
    a. prevents the bird from flapping its wings.
    b. creates more drag.
    c. decreases the amount of lift the bird can achieve.
    d. increases the amount of weight the bird must carry.

    ANS: C   DIF: II   OBJ: 3-3.2

63. Upside-down wings, or spoilers, mounted on the rear end of race cars help to reduce the danger of accidents because they
    a. decrease the air pressure above the car, decreasing lift.
    b. increase the air pressure above the car, decreasing lift.
    c. increase the air pressure below the car, decreasing lift.
    d. prevent the car from rolling.

    ANS: B   DIF: II   OBJ: 3-3.3

64. Meteorologists can predict the direction of a warm front by knowing the pressure of the front and the pressure around the front because
    a. a low-pressure front will move to an area with higher pressure.
    b. a low-pressure front will move to an area with the same pressure.
    c. a high-pressure front will move to an area with higher pressure.
    d. a high-pressure front will move to an area with lower pressure.

    ANS: D   DIF: I   OBJ: 3-1.3

65. You may feel discomfort in your ears when you take off in an airplane because the pressure in your inner ear
    a. is less than the atmospheric pressure at a high altitude.
    b. is greater than the atmospheric pressure at a high altitude.
    c. is equal to the atmospheric pressure at a high altitude.
    d. is not affected and cannot be related to the atmospheric pressure at a high altitude.

    ANS: B   DIF: II   OBJ: 3-1.2

66. A bicycle tire remains inflated because air particles
    a. form a crystal matrix inside the tire.
    b. are constantly moving and pushing against the inside of the tire.
    c. are less dense than the tire.
    d. displace the tire's density.

    ANS: B   DIF: I   OBJ: 3-1.1

67. A bicycle tire feels harder as you pump more air into it because as you pump more air into the tire,
    a. fewer air particles push against the walls of the tire, decreasing the tire pressure.
    b. fewer air particles push against the walls of the tire, increasing the tire pressure.
    c. more air particles push against the walls of the tire, decreasing the tire pressure.
    d. more air particles push against the walls of the tire, increasing the tire pressure.

    ANS: D   DIF: I   OBJ: 3-1.1

68. You do not notice atmospheric pressure on your body because
    a. it is air.
    b. atmospheric pressure acts evenly on your body.
    c. of gravity.
    d. the fluids inside your body also exert pressure.

    ANS: D   DIF: I   OBJ: 3-1.1

69. Astronauts wear pressurized suits in space because if they took their suits off,
    a. they would be unable to move.
    b. the fluids in their bodies would exert the same pressure as the pressure outside their bodies.
    c. the fluids in their bodies would overcome the pressure outside of their bodies, and they would explode.
    d. the fluids in their bodies would be overcome by the pressure outside of their bodies, and they would implode.

    ANS: C   DIF: III   OBJ: 3-1.2

70. The atmospheric pressure at an altitude of 12,000 m is about one-fifth that at sea level. Airplanes that fly at this altitude pressurize the cabins for passenger safety. If these cabins were not pressurized, passengers would feel uncomfortable because the fluids inside their bodies would exert a pressure
    a. one-fifth the pressure exerted on the outside of their bodies by the atmosphere.
    b. approximately five times the pressure exerted on the outside of their bodies by the atmosphere.
    c. would be the same as atmospheric pressure.
    d. would stop exerting pressure because of the high altitude.

    ANS: B   DIF: III   OBJ: 3-1.2

71. Air flows into your lungs as you inhale because the pressure inside your lungs
    a. is greater than that of the surrounding air, and air flows from lower to higher pressure.
    b. is less than that of the surrounding air, and air flows from higher to lower pressure.
    c. is the same as that of the surrounding air.
    d. does not change.

    ANS: B   DIF: I   OBJ: 3-1.3

72. Mountain climbers who trek to the summit of Mount Everest (8,847 m above sea level) have difficulty breathing because
    a. air stops flowing from high to low pressure.
    b. air stops flowing from low to high pressure.
    c. there is not that much difference in pressure between the inside of their lungs and the air.
    d. the difference between the pressure inside their lungs and the surrounding air is too great.

    ANS: C   DIF: II   OBJ: 3-1.3

73. When an athlete exhales during strenuous exercise, the lungs push out more air than if she were resting. When she inhales, the pressure inside her lungs _____ when she rests, resulting in deeper breaths.
    a. is greater than   c. is the same as
    b. is less than   d. is zero, unlike

    ANS: B   DIF: II   OBJ: 3-1.3

COMPLETION

1. _____ increases with the depth of a fluid. (Pressure or Lift)

   ANS: Pressure   DIF: I   OBJ: 3-1.2

2. A plane's engine produces _____ to push the plane forward. (thrust or drag)

   ANS: thrust   DIF: I   OBJ: 3-3.2

3. Force divided by area is known as _____. (density or pressure)

    ANS: pressure   DIF: I   OBJ: 3-1.1

4. The hydraulic brakes of a car transmit pressure through fluid. This is an example of
    _____. (Archimedes' principle or Pascal's principle)

    ANS: Pascal's principle   DIF: I   OBJ: 3-1.4

5. Bernoulli's principle states that the pressure exerted by a moving fluid is
    _____ the pressures of the fluid when it is not moving. (greater than or less than)

    ANS: less than   DIF: I   OBJ: 3-3.1

6. The fact that a heavy steel cargo ship can carry a large load without sinking illustrates
    _____. (Archimedes' principle or Bernoulli's principle)

    ANS: Archimedes' principle   DIF: I   OBJ: 3-2.2

7. The forward force produced by an airplane's engine is called _____. (lift or thrust)

    ANS: thrust   DIF: I   OBJ: 3-3.2

8. Hydraulic devices multiply force using _____. (Archimedes' principle or Pascal's principle)

    ANS: Pascal's principle   DIF: I   OBJ: 3-1.4

9. The _____ that a liquid exerts on an object increases as the density of the fluid increases. (buoyant force or lift)

    ANS: buoyant force   DIF: I   OBJ: 3-2.3

10. _____ works against the forward force of a plane. (Drag or Lift)

    ANS: Drag   DIF: I   OBJ: 3-3.2

11. Any material that can flow and that takes the shape of its container is a _____.

    ANS: fluid   DIF: I   OBJ: 3-1.1

12. One _____ is the force of one newton exerted over an area of one square meter.

    ANS: pascal   DIF: I   OBJ: 3-1.1

13. The pressure caused by the weight of the atmosphere is called _____.

ANS: atmospheric pressure   DIF: I   OBJ: 3-1.1

14. _____ is the amount of matter in a certain volume.

ANS: Density   DIF: I   OBJ: 3-1.2

15. _____ states that a change in pressure at any point in an enclosed fluid will be transmitted equally to all parts of the fluid.

ANS: Pascal's principle   DIF: I   OBJ: 3-1.4

16. The upward force that fluids exert on all matter is called _____.

ANS: buoyant force   DIF: I   OBJ: 3-2.1

17. _____ states that the buoyant force on an object in a fluid is an upward force equal to the weight of the volume of fluid that the object displaces.

ANS: Archimedes' principle   DIF: I   OBJ: 3-2.1

18. _____ states that as the speed of a moving fluid increases, its pressure decreases.

ANS: Bernoulli's principle   DIF: I   OBJ: 3-3.1

19. The faster-moving air above an airplane's wings exerts less pressure than the slower-moving air below its wings thereby creating an upward force that is called _____.

ANS: lift   DIF: I   OBJ: 3-3.2

20. The layer of oxygen, nitrogen, and other gases that surrounds the Earth is called the _____.

ANS: atmosphere   DIF: I   OBJ: 3-1.1

21. Devices that use liquids to transmit pressure from one point to another are called _____ devices.

ANS: hydraulic   DIF: I   OBJ: 3-1.4

22. Most bony fish have an organ called a swim bladder that allows a fish to adjust their overall _____ in order to go to certain depths in the water.

ANS: density   DIF: I   OBJ: 3-2.3

Holt Science and Technology
59

23. An irregular or unpredictable flow of fluids is known as _____.

    ANS: turbulence   DIF: I   OBJ: 3-3.2

24. Atmospheric and water pressure increase with depth because of _____.

    ANS: gravity   DIF: I   OBJ: 3-1.2

SHORT ANSWER

1. How do particles in a fluid exert pressure on a container?

   ANS:
   The moving particles in a fluid collide against each other and against the walls of the container.
   This creates pressure.

   DIF: I   OBJ: 3-1.1

2. Why are you not crushed by atmospheric pressure?

   ANS:
   The pressure exerted by the fluids in your body works against atmospheric pressure.

   DIF: I   OBJ: 3-1.2

3. Explain why dams on deep lakes should be thicker at the bottom than near the top.

   ANS:
   Water pressure increases with depth. Therefore, more pressure is exerted at the bottom of the
   dam than at the top. The dam must be thicker at the bottom to withstand this added pressure.

   DIF: II   OBJ: 3-1.2

4. Explain how atmospheric pressure helps you drink through a straw.

   ANS:
   When you sip through a straw, some of the air is removed from the straw. This reduces the
   pressure inside the straw. The pressure outside the straw is now greater than inside the straw.
   Because fluids flow from areas of high pressure to areas of low pressure, the liquid moves up in
   the straw.

   DIF: I   OBJ: 3-1.3

5. What does Pascal's principle state?

ANS:
Pascal's principle states that a change in pressure at any point in an enclosed fluid will be transmitted equally to all parts of that fluid.

DIF: I   OBJ: 3-1.4

6. When you squeeze a balloon, where is the pressure inside the balloon increased the most? Explain your answer in terms of Pascal's principle.

ANS:
The pressure inside the balloon is increased equally at all points because changes in fluid pressure are transmitted equally to all parts of an enclosed fluid.

DIF: II   OBJ: 3-1.4

7. Explain how differences in fluid pressure create buoyant force on an object.

ANS:
Water pressure is exerted on all sides of an object. The pressures exerted horizontally on both sides cancel each other out. The pressure exerted at the bottom is greater than that exerted at the top because pressure increases with depth. This creates an overall upward force on the object—the buoyant force.

DIF: I   OBJ: 3-2.1

8. An object weighs 20 N. It displaces a volume of water that weighs 15 N.
   a. What is the buoyant force on the object?
   b. Will this object float or sink? Explain your answer.

ANS:
   a. 15 N
   b. It will sink because its weight is greater than the buoyant force acting on it.

DIF: I   OBJ: 3-2.2

9. Iron has a density of 7.9 g/cm^3. Mercury has a density of 13.6 g/cm^3. Will iron float or sink in mercury? Explain your answer.

ANS:
It will float because it is less dense than mercury.

DIF: I   OBJ: 3-2.2

10. Why is it inaccurate to say that all heavy objects will sink in water?

ANS:
Whether an object sinks or floats depends on density, not weight. A heavy object can have an overall density that is less than that of water and can therefore float.

DIF: II   OBJ: 3-2.3

11. Does fluid pressure increase or decrease as fluid speed increases?

ANS:
Fluid pressure decreases.

DIF: I   OBJ: 3-3.1

12. Explain how wing shape can contribute to lift during flight.

ANS:
Many wings are designed so that air passing over the wing travels faster than air traveling under the wing. This faster-moving air reduces the pressure above the wing, and higher pressure below the wing results in lift (upward force on the wing).

DIF: I   OBJ: 3-3.2

13. What force opposes motion through a fluid?

ANS:
Drag is the force that opposes motion through a fluid.

DIF: I   OBJ: 3-3.1

14. When the space through which a fluid flows becomes narrow, fluid speed increases. Explain how this could lead to a collision for two boats passing very close to each other.

ANS:
As the fluid speed between the boats increases, the fluid pressure decreases. The pressure on the outer sides of the boats then becomes greater than the pressure between them. This increased pressure from the outside can push the boats together, causing them to collide.

DIF: II   OBJ: 3-3.3

15. What do liquids and gases have in common?

ANS:
They are both fluids.

DIF: I   OBJ: 3-1.1

16. Why does pressure increase with depth?

ANS:
As depth increases, the weight of the fluid above increases, which increases pressure.

DIF: I   OBJ: 3-1.2

17. What will happen to the pressure in all parts of an enclosed fluid if there is an increase in pressure in one part?

ANS:
According to Pascal's principle, a change in pressure at any point in an enclosed fluid will be transmitted equally to all parts of that fluid.

DIF: I   OBJ: 3-1.3

18. How can you determine the buoyant force acting on an object?

ANS:
Determine the weight of the volume of fluid displaced by the object.

DIF: I   OBJ: 3-2.1

19. What factors determine how heavy an object can be and still float?

ANS:
Density and shape determine how heavy an object can be and still float.

DIF: I   OBJ: 3-2.3

20. What happens when you place an object in water that has exactly the same density as water?

ANS:
It neither sinks nor floats; it remains where you place it.

DIF: I   OBJ: 3-2.3

21. How can a scuba diver keep from floating back to the surface of the water?

ANS:
The diver can add weights.

DIF: I   OBJ: 3-2.2

22. What forces act on an aircraft?

ANS:
Lift, thrust, drag, and gravity are forces that act on an aircraft.

DIF: I   OBJ: 3-3.2

23. When an airplane is flying, how does the air pressure above a wing compare with that below the wing?

ANS:
Air pressure above the wing is lower.

DIF: I   OBJ: 3-3.2

24. Why do shower curtains often have weights or magnets at the bottom?

ANS:
Shower curtains often have weights or magnets at the bottom in order to prevent them from being pushed toward the water stream.

DIF: I   OBJ: 3-3.3

25. What two factors determine the amount of lift achieved by an airplane?

ANS:
Thrust and wing size determine the amount of lift achieved by an airplane.

DIF: I   OBJ: 3-3.2

26. Where is water pressure greater, at a depth of 1 m in a large lake or at a depth of 2 m in a small pond? Explain.

ANS:
Water pressure is greater at a depth of 2 m in a small pond. Pressure increases with depth, regardless of the amount of fluid present.

DIF: I   OBJ: 3-1.2

27. Is there buoyant force on an object at the bottom of an ocean? Explain your reasoning.

ANS:
Yes; the object displaced fluid. The buoyant force on the object equals the weight of the water displaced. In this case, however, the weight of the object was larger than the buoyant force, so the object sank.

DIF: I   OBJ: 3-2.1

28. Why are liquids used in hydraulic brakes instead of gases?

ANS:
Liquids are used in hydraulic brakes because liquids cannot be compressed easily. Gases are compressible.

DIF: I   OBJ: 3-1.3

29. Use the following terms to create a concept map: *fluid, pressure, depth, buoyant force, density.*

ANS:

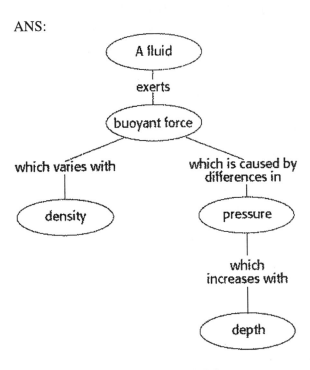

DIF: II   OBJ: 3-2.3

30. Compared with an empty ship, will a ship loaded with plastic-foam balls float higher or lower in the water? Explain your reasoning.

ANS:
The ship will float lower in the water because the plastic-foam balls will add to the total mass of the ship but will not increase the volume. Therefore, the overall density of the ship will increase, causing the ship to sink a little.

DIF: II   OBJ: 3-2.3

31. Inside all vacuum cleaners is a high-speed fan. Explain how this fan causes dirt to be picked up by the vacuum cleaner.

ANS:
The fan causes the air inside the vacuum cleaner to move faster, which decreases pressure. The higher air pressure outside of the vacuum then pushes dirt into the vacuum cleaner.

DIF: II   OBJ: 3-3.3

32. A 600 N clown on stilts says to two 600 N clowns sitting on the ground, "I am exerting twice as much pressure as the two of you together!" Could this statement be true? Explain your reasoning.

ANS:
Yes, the statement could be true. Pressure is equal to force over area, that is, an amount of force applied over a certain area. The clown on stilts is exerting force over a much smaller area than the two clowns on the ground are. Therefore, it is possible that the clown on stilts is exerting twice as much pressure as the other two clowns are.

DIF: II   OBJ: 3-1.1

33. Calculate the area of a 1,500 N object that exerts a pressure of 500 Pa (N/m^2). Then calculate the pressure exerted by the same object over twice that area. Be sure to express your answers in the correct SI unit.

ANS:
3 m^2; 250 Pa

DIF: II   OBJ: 3-1.1

Examine the illustration of an iceberg below and answer the questions that follow.

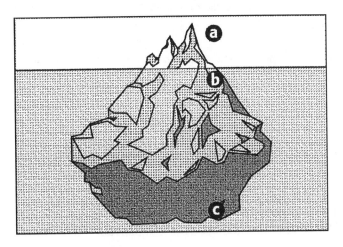

34. At what point (**a**, **b**, or **c**) is the water pressure greatest on the iceberg?

    ANS:
    c

    DIF: II   OBJ: 3-2.1

35. How much of the iceberg has a weight equal to the buoyant force?
    a. all of it
    b. the section from a to b
    c. the section from b to c

    ANS:
    a

    DIF: II   OBJ: 3-2.1

36. How does the density of ice compare with the density of water?

    ANS:
    Ice is less dense than water.

    DIF: II   OBJ: 3-2.3

37. Why do you think icebergs are so dangerous to passing ships?

    ANS:
    Only a small portion of an iceberg floats above water, as shown in the image. A ship may
    actually be closer to running into a massive block of ice underwater than it would appear on the
    surface. If the ship is not turned or stopped in time, it could collide with or scrape the iceberg.

    DIF: II   OBJ: 3-2.1

38. As an airplane comes in for a landing, the pilot often raises the wing flaps. For what purpose other than steering the plane would the wing flaps be raised? Explain.

ANS:
The pilot must slow down the airplane when landing. The raised wing flaps create turbulence. The turbulence creates drag, which slows down the plane.

DIF: I   OBJ: 3-3.2

39. Wood usually floats, but some types of wood are more buoyant than others. For example, a cubic meter of balsa wood is more buoyant than a cubic meter of oak. Explain why this is true.

ANS:
The cubes have equal volumes, but the cube of oak has more mass per unit volume—or a greater density—than the cube of balsa wood. Because the cube of oak is less buoyant, more of the cube is under the water.

DIF: I   OBJ: 3-2.3

40. Why does water pressure increase as the depth of the water increases?

ANS:
The weight of the atmosphere plus the weight of the water above presses down due to gravity. At greater depths, more water presses down from above, increasing water pressure.

DIF: I   OBJ: 3-1.2

41. The human heart contains valves that direct the flow of blood. Healthy valves offer very little resistance to the forward flow of blood. When blood passes through a valve that has been narrowed by disease, a sound called a *murmur* can be heard. What fluid principle or action do you think might produce this sound?

ANS:
Blood passing through a narrowed heart valve encounters resistance, which causes turbulent blood flow. This turbulence causes vibrations that are transmitted to the surface of the chest. These vibrations are heard as a heart murmur.

DIF: II   OBJ: 3-1.3

42. The following graph shows the relationship between atmospheric pressure and altitude above sea level. Examine the graph and answer the following question.

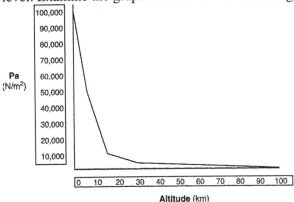

a. In which 10 km interval does air pressure change the most?
b. What conclusions can you draw from this graph?

ANS:
a. Air pressure changes the most in the first 10 km interval.
b. Sample answer: As you move higher into the atmosphere, the atmospheric pressure decreases. Above 30 km, atmospheric pressure does not change as much.

DIF: II   OBJ: 3-1.2

43. The pressure exerted on a submarine 10 m below the surface of the water is 100 kPa. How deep is the submarine if the water pressure is 5,000 kPa? Show your work.

ANS:
100 kPa ÷ 10 m = 10 kPa ÷ 1 m
5,000 kPa ÷ (10 kPa ÷ 1 m) = 500 m

DIF: II   OBJ: 3-2.1

44. Use the following terms to complete the concept map below: *decreases, Bernoulli's principle, multiplies, Pascal's principle.*

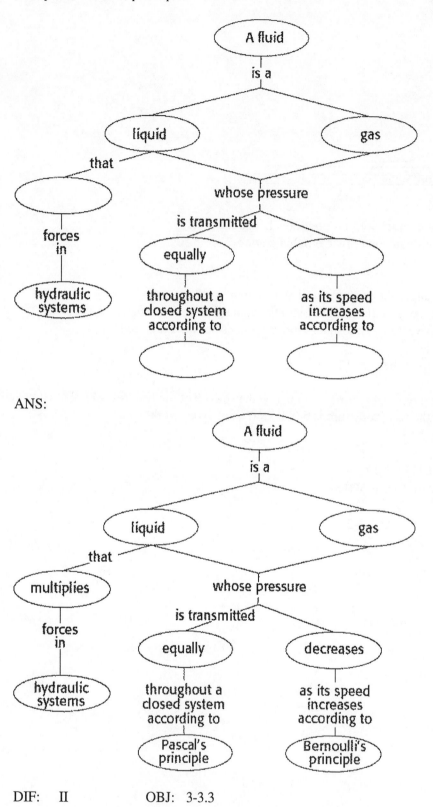

ANS:

DIF: II   OBJ: 3-3.3

MULTIPLE CHOICE

1. Work is being done when
   a. you apply a force to an object.
   b. an object is moving after you apply a force to it.
   c. you exert a force that moves an object in the direction of the force.
   d. you do something that is difficult.

   ANS: C   DIF: I   OBJ: 4-1.1

2. The work output for a machine is always less than the work input because
   a. all machines have a mechanical advantage.
   b. some of the work done is used to overcome friction.
   c. some of the work done is used to overcome distance.
   d. power is the rate at which work is done.

   ANS: B   DIF: I   OBJ: 4-2.2

3. The unit for work is the
   a. joule.
   b. joule per second.
   c. newton.
   d. watt.

   ANS: A   DIF: I   OBJ: 4-1.1

4. Which of the following is NOT a simple machine?
   a. a faucet handle
   b. a jar lid
   c. a can opener
   d. a seesaw

   ANS: C   DIF: I   OBJ: 4-3.3

5. Power is
   a. how strong someone or something is.
   b. how much force is being used.
   c. how much work is being done.
   d. how fast work is being done.

   ANS: D   DIF: I   OBJ: 4-1.3

6. The unit for power is the
   a. newton.
   b. kilogram.
   c. watt.
   d. joule.

   ANS: C   DIF: I   OBJ: 4-1.3

7. A machine can increase
   a. distance at the expense of force.
   b. force at the expense of distance.
   c. neither distance nor force.
   d. Both (a) and (b)

   ANS: D   DIF: I   OBJ: 4-2.1

8. In which situation is a person doing work on an object?
   a. A school crossing guard raises a stop sign that weighs 10 N.
   b. A student walks 1 m/s while wearing a backpack that weighs 15 N.
   c. A man exerts a 350 N force on a rope attached to a house.
   d. A worker holds a box 1 m off the floor.

   ANS: A   DIF: I   OBJ: 4-1.1

9. Juan and Anita each lift an identical stack of books the same distance onto a table, but Anita does the job twice as fast. Therefore, her actions involve twice as much
   a. work input.   c. power.
   b. work output.   d. efficiency.

   ANS: C   DIF: I   OBJ: 4-1.3

10. Which of the following machines always has a mechanical advantage of less than 1?
    a. wheel and axle   c. a long, thin wedge
    b. third-class lever   d. a poorly lubricated, movable pulley

    ANS: B   DIF: I   OBJ: 4-2.3

11. What does a fixed pulley change?
    a. both the size and direction of a force
    b. the size of a force
    c. the direction of a force
    d. neither the direction nor the size of a force

    ANS: C   DIF: I   OBJ: 4-3.1

12. When a machine increases the size of the force exerted, the distance through which the force is exerted
    a. must increase.   c. must stay the same.
    b. must decrease.   d. must double.

    ANS: B   DIF: I   OBJ: 4-2.2

13. Which of the following would increase the mechanical advantage of a first-class lever?
    a. applying a greater input force
    b. decreasing the load
    c. increasing the rate at which force is applied
    d. moving the fulcrum closer to the load

    ANS: D   DIF: I   OBJ: 4-3.2

14. Levers are divided into classes according to the location of
    a. the fulcrum.   c. the input force.
    b. the load.   d. the fulcrum, the load, and the input force.

    ANS: D   DIF: I   OBJ: 4-3.1

15. Which of the following statements about inclined planes is NOT true?
    a. Inclined planes allow you to apply a smaller force over a smaller distance.
    b. Egyptians used inclined planes to build the Great Pyramid.
    c. An example of an inclined plane is a wedge.
    d. A screw is a type of inclined plane.

    ANS: A   DIF: I   OBJ: 4-3.1

16. Which of the following is NOT an example of work according to the scientific definition of work?
    a. playing baseball
    b. reading a chapter for homework
    c. pushing a wheelbarrow
    d. bowling

    ANS: B   DIF: I   OBJ: 4-1.1

17. A man applies a force of 500 N to push a truck 10 m down the street. How much work has been done?
    a. 50 J
    b. 500 J
    c. 510 J
    d. 5,000 J

    ANS: D   DIF: II   OBJ: 4-1.2

18. In a basketball game, what is NOT considered to be work done on an object?
    a. dribbling the ball in place
    b. slam-dunking the ball
    c. running down the court without the ball
    d. throwing the ball to a teammate

    ANS: C   DIF: I   OBJ: 4-1.1

19. Which of the following actions do more work on an object?
    a. lifting an 80 N box 1 m up off the floor
    b. lifting a 160 N box 1 m up off the floor
    c. lifting a 90 N box 2 m up off the floor
    d. lifting a 100 N box 1.5 m up off the floor

    ANS: C   DIF: II   OBJ: 4-1.2

20. You increase the amount of work being done on a box when you
    a. increase the distance to where you are pushing it.
    b. increase the speed at which you are pushing it.
    c. get it stuck halfway to the place where you are pushing it.
    d. pick it up halfway there and carry it the rest of the way.

    ANS: A   DIF: I   OBJ: 4-1.1

21. If you do 50 J of work in 5 s, your power is
    a. 10 W.
    b. 45 W.
    c. 55 W.
    d. 250 W.

    ANS: A   DIF: II   OBJ: 4-1.3

22. Which of the following is NOT a machine?
    a. a pair of scissors
    b. a glass
    c. a screw
    d. a bottle opener

    ANS: B   DIF: I   OBJ: 4-2.1

23. Input force
    a. is always equal to the weight of the object you are moving.
    b. is work done by the machine after you have applied a force to the machine.
    c. changes in size or direction when done by a machine.
    d. opposes the forces you and the machine are working against.

    ANS: C   DIF: I   OBJ: 4-2.1

24. You apply 200 N to a machine and the machine applies 2,000 N to an object. What is the mechanical advantage?
    a. $\frac{1}{10}$
    b. 10
    c. 1,800
    d. 400,000

    ANS: B   DIF: II   OBJ: 4-2.3

25. You apply 10 N to a machine and the machine applies 10 N to move another object. Since the input force is equal to the output force,
    a. this machine is not useful.
    b. this machine moved the object a great distance.
    c. the mechanical advantage is 1.
    d. this machine changed the input force.

    ANS: C   DIF: I   OBJ: 4-2.3

26. Which of the following could be a machine's mechanical efficiency?
    a. 95 N•m
    b. 95 W
    c. 95 J/s
    d. 95%

    ANS: D   DIF: I   OBJ: 4-2.4

27. Getting regular oil changes for your car improves the car's
    a. mechanical efficiency.
    b. work input.
    c. mechanical advantage.
    d. input force.

    ANS: A   DIF: I   OBJ: 4-2.4

28. A ramp is an example of which simple machine?
    a. lever
    b. inclined plane
    c. wheel and axle
    d. pulley

    ANS: B   DIF: I   OBJ: 4-3.1

29. Your muscles and bones form a(n)
    a. lever.
    b. inclined plane.
    c. wheel and axle.
    d. pulley.

    ANS: A   DIF: I   OBJ: 4-3.1

30. In a rowboat, the fulcrum (the place where the oars are in contact with the boat) is closer to the input force (your hand pulling and pushing on the oars) than to the load (the water) in order to
    a. decrease the output force.
    b. increase the output force.
    c. decrease the distance.
    d. increase the distance.

    ANS: D   DIF: I   OBJ: 4-3.2

31. Third-class levers
    a. do not change the direction of the input force.
    b. do not increase the input force.
    c. result in an output force less than the input force.
    d. All of the above

    ANS: D   DIF: I   OBJ: 4-3.2

32. What is the mechanical advantage of an inclined plane 3 m long and 0.5 m high?
    a. 0.5
    b. 1.5
    c. 3
    d. 6

    ANS: D   DIF: II   OBJ: 4-3.2

33. Which of the following is NOT a wedge?
    a. chisel
    b. axe head
    c. bottle opener
    d. knife

    ANS: C   DIF: I   OBJ: 4-3.1

34. Which knife gives you a better mechanical advantage?
    a. a short and fat knife
    b. a short and thin knife
    c. a long and fat knife
    d. a long and thin knife

    ANS: D   DIF: I   OBJ: 4-3.2

35. A jar lid is an example of a
    a. lever.
    b. screw.
    c. wheel and axle.
    d. wedge.

    ANS: B   DIF: I   OBJ: 4-3.1

36. A screw has the greatest mechanical advantage when it has a
    a. long spiral and threads that are tightly coiled.
    b. long spiral and threads that are loosely coiled.
    c. short spiral and threads that are tightly coiled.
    d. short spiral and threads that are loosely coiled.

    ANS: A   DIF: I   OBJ: 4-3.2

37. Which of the following is an example of a wheel and axle?
   a. doorknob
   b. wrench
   c. Ferris wheel
   d. all of the above

   ANS: D   DIF: I   OBJ: 4-3.2

38. What is the mechanical advantage of a wheel and axle where the wheel's radius is 10 cm and the axle's radius is 2 cm?
   a. 2
   b. 5
   c. 10
   d. 20

   ANS: B   DIF: II   OBJ: 4-3.2

39. With a fixed pulley, the input force
   a. is always greater than the output force.
   b. is always less than the output force.
   c. is always equal to the output force.
   d. opposes the output force.

   ANS: C   DIF: I   OBJ: 4-3.2

40. Which of the following has the greatest mechanical advantage?
   a. a fixed pulley
   b. a movable pulley with 2 segments
   c. a block and tackle with 4 segments
   d. a plane 3 m long and 1 m high

   ANS: C   DIF: II   OBJ: 4-3.2

41. A zipper is made from three
   a. inclined planes.
   b. wedges.
   c. levers.
   d. screws.

   ANS: B   DIF: I   OBJ: 4-3.3

42. Mechanical efficiency of compound machines is often
   a. quite high.
   b. 100%.
   c. quite low.
   d. nonexistent.

   ANS: C   DIF: I   OBJ: 4-3.3

43. A wedge that has a mechanical advantage of 4 and a length of 8 cm must be
   a. 2 cm wide.
   b. 4 cm wide.
   c. 12 cm wide.
   d. 32 cm wide.

   ANS: A   DIF: II   OBJ: 4-3.2

44. When you turn a screw, you apply a
   a. small input force over a long distance.
   b. large input force over a short distance.
   c. small input force over a short distance.
   d. large input force over a long distance.

   ANS: A   DIF: I   OBJ: 4-3.2

45. When a small input force is applied to the wheel and axle, it rotates through a
    a. short linear distance.
    b. long linear distance.
    c. circular distance.
    d. fulcrum.

    ANS: C   DIF: I   OBJ: 4-3.2

46. Which simple machines are known to have been used by the Egyptians to move and stack the huge blocks that form the Great Pyramids?
    a. wheel and axle
    b. lever and inclined plane
    c. wedge and screw
    d. pulley and block and tackle

    ANS: B   DIF: I   OBJ: 4-3.1

47. Which of the following is NOT an example of doing work on a suitcase?
    a. carrying a heavy suitcase down a long hallway
    b. lifting a heavy suitcase up and putting it on a top shelf
    c. pulling a wheeled suitcase down a long corridor
    d. sliding a heavy suitcase across a bed to move it out of the way

    ANS: A   DIF: I   OBJ: 4-1.1

48. Which of the following has the greatest mechanical advantage?
    a. a machine to which you apply a force of 50 N and the machine applies a force of 150 N
    b. a machine to which you apply a force of 60 N and the machine applies a force of 200 N
    c. a machine to which you apply a force of 25 N and the machine applies a force of 100 N
    d. a machine to which you apply a force of 40 N and the machine applies a force of 125 N

    ANS: C   DIF: II   OBJ: 4-2.3

49. Which of the following machines is the MOST efficient?
    a. a machine in which you input 4 J of work and it outputs 3 J of work
    b. a machine in which you input 12 J of work and it outputs 6 J of work
    c. a machine in which you input 8 J of work and it outputs 7 J of work
    d. a machine in which you input 10 J of work and it outputs 9 J of work

    ANS: D   DIF: II   OBJ: 4-2.4

50. You apply an input force of 20 N to a machine and it gives an output force of 100 N. What is this machine's mechanical advantage?
    a. $\frac{1}{5}$
    b. $\frac{1}{2}$
    c. 5
    d. 10

    ANS: C   DIF: II   OBJ: 4-2.3

51. If a machine has a work input of 8 N and gives work output of 6 N, what is this machine's mechanical efficiency?
    a. 5%
    b. 20%
    c. 75%
    d. 133%

    ANS: C   DIF: II   OBJ: 4-2.4

52. When you move a lever's fulcrum closer to the load than to the input force, you achieve
    a. a mechanical advantage less than 1.
    b. a greater output force over a shorter distance.
    c. a lower output force over a longer distance.
    d. the same amount of force as before.

    ANS: B   DIF: I   OBJ: 4-3.2

53. Which of the following is an example of a second-class lever?
    a. hammer   c. wheelbarrow
    b. seesaw   d. oars on a boat

    ANS: C   DIF: I   OBJ: 4-3.1

54. In a second-class lever,
    a. the load is between the fulcrum and the input force.
    b. the fulcrum is between the input force and the load.
    c. the input force is between the fulcrum and the load.
    d. the fulcrum is between the input force and the output force.

    ANS: A   DIF: I   OBJ: 4-3.1

55. Hammering a nail into wood is an example of using a
    a. wedge.   c. second-class lever.
    b. first-class lever   d. third-class lever.

    ANS: D   DIF: I   OBJ: 4-3.1

56. In the simple machine of a wheel and axle, a small input force is applied to a wheel. As the
    wheel turns, so does the axle. Because the axle is smaller than the wheel, it rotates through a
    a. smaller distance, which makes the output force smaller than the input force.
    b. smaller distance, which makes the output force larger than the input force.
    c. larger distance, which makes the output force larger than the input force.
    d. larger distance, which makes the output force smaller than the input force.

    ANS: B   DIF: I   OBJ: 4-3.2

57. A movable pulley
    a. is attached to the load.   c. has a mechanical advantage of 1.
    b. works by pulling down on the rope.   d. has the same output force as input force.

    ANS: A   DIF: I   OBJ: 4-3.1

58. Which of the following light bulbs does the most work each second?
    a. a 40 W bulb
    b. a 60 W bulb
    c. a 100 W bulb
    d. All bulbs do the same amount of work each second.

    ANS: C   DIF: II   OBJ: 4-1.3

59. When a machine changes the size of a force, the ____ must also change.
    a. work
    b. force
    c. distance
    d. direction

    ANS: C   DIF: I   OBJ: 4-2.2

60. You are loading the back of a truck with your new drum set. Which of the following does the LEAST amount of work on the drum set?
    a. lifting the drum set with your brute strength and setting it in the truck
    b. using a ramp to slide the drum set into the truck
    c. using a pulley to raise the drum set onto the truck
    d. All of these situations have the same amount of work being done on the drum set.

    ANS: D   DIF: I   OBJ: 4-2.2

61. To do the same amount of work on an object, a large force applied over a short distance can be exchanged by a
    a. small force over a short distance.
    b. small force over a long distance.
    c. large force over a long distance.
    d. None of the above

    ANS: B   DIF: I   OBJ: 4-2.2

62. A single pulley changes the ____ an object.
    a. direction of the input force on
    b. work done on
    c. amount of force exerted on
    d. distance that you will move

    ANS: A   DIF: I   OBJ: 4-2.2

63. Suppose you live on the top floor of the Lake Point Tower in Chicago, 195 m above the ground. If your weight is 500 N, what minimum amount of work must the elevator do to lift you from the ground floor to the top floor?
    a. 305 J
    b. 97,500 J
    c. 695 J
    d. 2.5 J

    ANS: B   DIF: II   OBJ: 4-1.2

64. In 1989, a hamburger with a mass of 2,500 kg was cooked in Wisconsin. What is the mechanical advantage of the ramp used to move the hamburger if the ramp has a length of 100 m and a height of 20 m?
    a. 125
    b. 100
    c. 5
    d. 0.2

    ANS: C   DIF: I   OBJ: 4-3.2

65. In 1994, a 3,000 kg pancake was cooked, and flipped, in Manchester, England. Suppose you had a giant spatula that you could use as a lever to flip this pancake. Where could you best position the fulcrum and the pancake in order to get the greatest mechanical advantage?
    a. place the fulcrum close to the pancake, as in a first-class lever
    b. place the pancake between the input force and the load, as in a second-class lever
    c. position the input force between the fulcrum and the pancake, as in a third-class lever
    d. All of these machines have equal mechanical advantage.

    ANS: A   DIF: I   OBJ: 4-3.2

66. The largest palace in the world is the Imperial Palace in Beijing, China. The palace covers a rectangle 750 m long by 960 m wide. If you were to push a lawn mower around the perimeter, applying a force of 60 N, how much work would you do?
    a. 43,200,000 J   c. 102,600 J
    b. 205,200 J   d. 45,000 J

    ANS: B   DIF: III   OBJ: 4-1.2

67. The largest turtle ever caught in the United States weighed about 844 N. How much work would have to be done to lift this turtle slowly 2 m from the surface of the water onto the deck of a research ship?
    a. 422 J   c. 846 J
    b. 842 J   d. 1688 J

    ANS: D   DIF: I   OBJ: 4-1.2

68. In 1453, during the siege of Constantinople, the Turks used a cannon capable of launching a stone cannonball with a mass of 540 kg. Suppose a soldier dropped a cannonball with this mass while trying to load it into the cannon. If the cannon ball fell 1 m before it landed on the soldier's toe, how much work was done by gravity on the cannonball? (Hint: The gravitational acceleration acting on the cannonball is 9.8 m/s/s.)
    a. 5292 J   c. 540 J
    b. 549.8 J   d. 9.8 J

    ANS: A   DIF: III   OBJ: 4-1.2

69. The longest shish kebab ever made was 881 m long. Suppose meat and vegetables need to be delivered in a crate from one end of this shish kebab's skewer to the other end. A cook pushes the crate by applying a net force of 40 N. What is the work done on the crate and its contents by the cook during the delivery?
    a. 35,240 J   c. 22 J
    b. 841 J   d. 0 J

    ANS: A   DIF: I   OBJ: 4-1.2

70. Which of the following levers provides a mechanical advantage of greater than 1?
    a. first-class lever   c. third-class lever
    b. second-class lever   d. Both (a) and (b)

    ANS: D   DIF: I   OBJ: 4-3.2

71. Which of the following levers provides a mechanical advantage of less than 1?
    a. first-class lever
    b. second-class lever
    c. third-class lever
    d. Both (a) and (b)

    ANS: C   DIF: I   OBJ: 4-3.2

COMPLETION

1. A _____ is the SI unit equivalent to 1 N•m. (watt or joule)

    ANS: joule   DIF: I   OBJ: 4-1.1

2. The work you do on a machine, such as turning a screwdriver, is called _____ (work input or work output)

    ANS: work input   DIF: I   OBJ: 4-2.1

3. Because of friction, the _____ of a machine is always less than 100 percent. (mechanical advantage or mechanical efficiency)

    ANS: mechanical efficiency   DIF: I   OBJ: 4-2.4

4. A _____ is a bar that pivots on a fulcrum. (lever or wedge)

    ANS: lever   DIF: I   OBJ: 4-3.1

5. A block and tackle is an example of a _____. (wheel and axle or compound machine)

    ANS: compound machine   DIF: I   OBJ: 4-3.3

6. _____ is done on an object when a force exerted on the object causes it to move in the direction of the force.

    ANS: Work   DIF: I   OBJ: 4-1.1

7. A _____ is also expressed as joules per second (J/s).

    ANS: watt   DIF: I   OBJ: 4-1.2

8. A _____ is a device that changes the size or direction of the force exerted on an object.

    ANS: machine   DIF: I   OBJ: 4-2.1

9. Work done by a machine on an object is called _____.

    ANS: work output   DIF: I   OBJ: 4-2.1

10. _____ is the force a machine applies through a distance.

    ANS: Work output   DIF: I   OBJ: 4-2.1

11. A machine's _____ indicates how many times the machine multiplies force by comparing the input force with the output force.

    ANS: mechanical advantage   DIF: I   OBJ: 4-2.3

12. A machine's _____ is a comparison of a machine's work output with the work input.

    ANS: mechanical efficiency   DIF: I   OBJ: 4-2.4

13. An _____ is a simple machine that is a straight, slanted surface.

    ANS: inclined plane   DIF: I   OBJ: 4-3.1

14. A _____ is a double inclined plane that moves.

    ANS: wedge   DIF: I   OBJ: 4-3.1

15. A _____ is an inclined plane that is wrapped in a spiral.

    ANS: screw   DIF: I   OBJ: 4-3.1

16. A _____ is a simple machine consisting of two circular objects of different sizes.

    ANS: wheel and axle   DIF: I   OBJ: 4-3.1

17. A _____ is a simple machine consisting of a grooved wheel that holds a rope or a cable.

    ANS: pulley   DIF: I   OBJ: 4-3.1

18. _____ are machines consisting of two or more simple machines.

    ANS: Compound machines   DIF: I   OBJ: 4-3.3

19. A pulley system in which a fixed pulley and a movable pulley are used together is called a _____.

    ANS: block and tackle   DIF: I   OBJ: 4-3.3

20. _____ is the distance from the center to the edge of a circle.

    ANS: Radius   DIF: I   OBJ: 4-3.1

21. A _____ pulley is attached to something that does NOT move.

ANS: fixed   DIF: I   OBJ: 4-3.1

22. A _____ pulley moves up with the load as it is lifted.

ANS: movable   DIF: I   OBJ: 4-3.1

23. The less work a machine has to do to overcome friction, the more _____ it is.

ANS: efficient   DIF: I   OBJ: 4-2.4

24. A _____ is a fixed point at which a lever pivots.

ANS: fulcrum   DIF: I   OBJ: 4-3.1

SHORT ANSWER

For each pair of terms, explain the differences in their meanings.

1. joule/watt

ANS:
A *joule* is a unit used to express work. A *watt* is a unit used to express power.

DIF: I   OBJ: 4-1.3

2. work output/work input

ANS:
Work output—work done by a machine; *work input*—work you do on a machine.
(Work output is always less than work input.)

DIF: I   OBJ: 4-2.1

3. mechanical efficiency/mechanical advantage

ANS:
Mechanical efficiency and mechanical advantage both compare characteristics of a machine.
Mechanical efficiency is a percentage that compares a machine's work output with its work
input. *Mechanical advantage* compares a machine's output force with its input force.

DIF: I   OBJ: 4-2.4

4. screw/inclined plane

   ANS:
   An inclined plane and a screw are both simple machines. An *inclined plane* is a straight, slanted surface, such as a ramp. A *screw* is an inclined plane wrapped in a spiral.

   DIF: I   OBJ: 4-3.1

5. simple machine/compound machine

   ANS:
   A *simple machine* is one of the following: inclined plane, wedge, screw, wheel and axle, pulley, or lever. A *compound machine* consists of one or more simple machines.

   DIF: I   OBJ: 4-3.3

6. Work is done on a ball when a pitcher throws it. Is the pitcher still doing work on the ball as it flies through the air? Explain.

   ANS:
   The pitcher is no longer doing work on the ball as it flies through the air because he is no longer exerting a force on it. (However, work is being done on the ball by the Earth, which exerts a force on the ball and pulls it back toward the ground.)

   DIF: I   OBJ: 4-1.1

7. Explain the difference between work and power.

   ANS:
   Work occurs when a force causes an object to move in the direction of the force, and power is the rate at which work is done. The more work you do in a given amount of time, or the less time it takes you to do a given amount of work, the greater your power.

   DIF: I   OBJ: 4-1.3

8. You lift a chair that weighs 50 N to a height of 0.5 m and carry it 10 m across the room. How much work do you do on the chair?

   ANS:
   Work is done on the chair only when it is picked up, not when it is carried across the room. Therefore, $W = 50 \text{ N} \times 0.5 \text{ m} = 25 \text{ J}$.

   DIF: II   OBJ: 4-1.2

9. Explain how using a ramp makes work easier.

    ANS:
    Sample answer: Using a ramp makes work easier because it allows you to apply a smaller input force than you would have to apply when lifting a load straight up. However, the smaller force has to be exerted over a longer distance.

    DIF: I   OBJ: 4-2.2

10. Why is it impossible for a machine to be 100 percent efficient?

    ANS:
    Sample answer: A machine can't be 100 percent efficient because some of the work input is used to overcome friction. Therefore, work input is always greater than work output.

    DIF: I   OBJ: 4-2.4

11. Suppose you exert 15 N on a machine, and the machine exerts 300 N on another object. What is the machine's mechanical advantage?

    ANS:
    $MA = 300 \text{ N} \div 15 \text{ N} = 20$

    DIF: I   OBJ: 4-2.3

12. Suppose you exert 15 N on a machine, and the machine exerts 300 N on another object. How does the distance through which the output force is exerted differ from the distance through which the input force is exerted?

    ANS:
    Because the output force is greater than the input force, the distance over which the output force is exerted must be shorter than the distance over which the input force is exerted. This is an example of the force-distance trade-off.

    DIF: II   OBJ: 4-2.2

13. Give an example of each of the following simple machines: first-class lever, second-class lever, third-class lever, inclined plane, wedge, and screw.

    ANS:
    Sample answers: *first-class lever:* a screwdriver used to pry the lid off a paint can; *second-class lever:* a wheelbarrow; *third-class lever:* your leg as you kick a soccer ball (your knee is the fulcrum); *inclined plane:* a ramp on the back o f a moving truck; *wedge:* a knife; *screw:* a jar lid

    DIF: I   OBJ: 4-3.1

14. A third-class lever has a mechanical advantage of less than 1. Explain why it is useful for some tasks.

ANS:
A third-class lever helps because it increases the distance through which the output force is exerted. For example, when you move the handle of a fishing pole just slightly, the other end of the pole moves a great distance.

DIF: I   OBJ: 4-3.2

15. Give an example of a wheel and an axle.

ANS:
Examples include the crank on a can opener, the reel on a fishing rod, a screwdriver, a doorknob, the crank on an ice cream maker, and the film-advance mechanism on an old camera.

DIF: I   OBJ: 4-3.3

16. Identify the simple machines that make up tweezers and nail clippers.

ANS:
Each side of the tweezers is a third-class lever. The sharpened edges of nail clippers are wedges, and the arm that activates the clipper is a second-class lever.

DIF: I   OBJ: 4-3.3

17. The radius of the wheel of a wheel and axle is four times greater than the radius of the axle. What is the mechanical advantage of this machine?

ANS:
The mechanical advantage of a wheel and axle is determined by the ratio of the wheel radius to the axle radius. So this machine would have a mechanical advantage of 4.

DIF: II   OBJ: 4-3.2

18. What are the two things that must happen for work to be done?

ANS:
A force must be exerted on an object, and the object must move in the direction of the force.

DIF: I   OBJ: 4-1.1

19. You push a 75 N box 3 m across the floor. How much work has been done?

ANS:
A total of 225 J of work has been done.

DIF: I   OBJ: 4-1.2

20. What is the power of a small motor that can do 4500 J of work in 25 s?

ANS:
This motor has 180 W of power.

DIF: I   OBJ: 4-1.3

21. How does a machine make work easier?

ANS:
A machine makes work easier by changing the size or direction, or both, of a force.

DIF: I   OBJ: 4-2.1

22. What two things do you need to know in order to calculate mechanical efficiency?

ANS:
Work input and work output must be known in order to calculate mechanical efficiency.

DIF: I   OBJ: 4-2.4

23. If the mechanical advantage of a machine is 5, how does the output force compare with the input force?

ANS:
The output force is five times greater than the input force.

DIF: I   OBJ: 4-2.3

24. Why are simple machines so useful?

ANS:
They make work easier.

DIF: I   OBJ: 4-3.2

25. Identify types of simple machines you might find on a playground. Describe how each of them modifies work.

ANS:
Answers may vary. Possible answers: seesaw—lever changes direction of input force; merry-go-round—wheel and axle makes the input force on the axle cause the wheel to move in a circle

DIF: I   OBJ: 4-3.1

26. How does reducing friction increase the mechanical efficiency of a compound machine?

ANS:
Less work input is used to overcome friction, so work output is higher and mechanical efficiency is higher.

DIF: I   OBJ: 4-3.3

27. Identify the simple machines that make up a pair of scissors.

ANS:
A pair of scissors consists of two first-class levers (arms) and two wedges (blades).

DIF: I   OBJ: 4-3.3

28. Explain the force-distance trade-off that occurs when a machine is used to make work easier.

ANS:
Sample answer: Some machines allow you to apply a small force over a large distance in order to get a larger output force. That larger output force, however, is exerted over a smaller distance. Sometimes it is preferable for a machine to increase distance, as when using chopsticks as third-class levers. Chopsticks allow you to move your fingers a little distance to move the ends of the chopsticks a larger distance.

DIF: I   OBJ: 4-2.2

29. Explain why you do work on a bag of groceries when you pick it up but NOT when you carry it.

ANS:
Sample answer: Work is done when motion due to a force occurs in the same direction as the force. When you pick up a bag of groceries, you exert a force up, and the bag moves up, so you are doing work. When you carry a bag of groceries, you exert an upward force but you are moving the bag forward, so you are not doing work.

DIF: I   OBJ: 4-1.1

30. Use the following terms to create a concept map: *work, force, distance, machine, mechanical advantage.*

ANS:

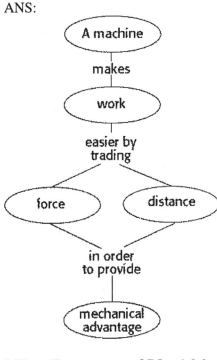

DIF: II   OBJ: 4-2.3

31. Why do you think levers usually have a greater mechanical efficiency than other simple machines do?

ANS:
Sample answer: Because some work input is used to overcome friction, work output is always less than work input. Levers do not have a lot of moving parts, so they don't generate as much friction as other machines. Less work input is used to overcome friction. As a result, the mechanical efficiency of a lever is usually greater than that of other simple machines.

DIF: II   OBJ: 4-2.4

32. A winding road is actually a series of inclined planes. Describe how a winding road makes it easier for vehicles to travel up a hill.

ANS:
Sample answer: A winding road makes climbing a hill easier because the length of the winding road is longer than the length of a road straight up the hill. Because the mechanical advantage of an inclined plane is determined by dividing the length of the inclined plane by its height, the more the road winds, the easier it is for a car to get up the hill.

DIF: II   OBJ: 4-2.2

33. Why do you think you would not want to reduce the friction involved in using a winding road?

ANS:
Sample answer: Friction between the road and the car tires is necessary for a car to travel along the road. Although friction reduces mechanical efficiency, reducing the friction between tires and the roadway would prevent cars from traveling safely along a winding road.

DIF: II   OBJ: 4-2.4

34. You and a friend together apply a force of 1,000 N to a 3,000 N automobile to make it roll 10 m in 1 min and 40 s.
    a. How much work did you and your friend do together?
    b. What was your combined power?

ANS:

    a. *Work* = 1,000 N × 10 m = 10,000 J
    b. *Power* = 10,000 J ÷ 100 s = 100 W

DIF: II   OBJ: 4-1.3

For each of the images below, identify the class of lever used and calculate the mechanical advantage.

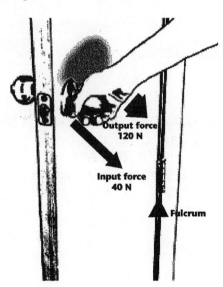

35.

ANS:
second-class lever, MA = 120 N ÷ 40 N = 3

DIF: II   OBJ: 4-3.2

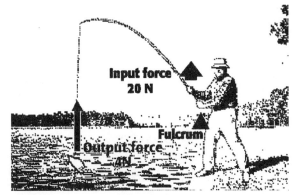

Input force 20 N

Fulcrum

Output force 4N

36.

ANS:
third-class lever, MA = 4 N ÷ 20 N = 0.20

DIF: II   OBJ: 4-3.2

37. If a person holds a rock over his or her head, how much work is being performed on the rock?
    Explain.

ANS:
Work occurs when an object moves in the same direction as the force that is applied to it.
Holding the rock does not involve any work because the rock is not being moved by force.

DIF: I   OBJ: 4-1.1

38. Explain why a pair of scissors is a compound machine.

ANS:
A pair of scissors is made of two simple machines. The blades of scissors are wedges. The blades
are also two first-class levers that are fastened together by and pivot on a fulcrum.

DIF: I   OBJ: 4-3.3

39. Explain why work output can never be equal to or greater than work input.

ANS:
Work input is always greater than work output because a portion of the work input is used to
overcome friction created by the machine.

DIF: I   OBJ: 4-2.4

40. Use the following terms to complete the concept map below: *mechanical efficiency, length, height, mechanical advantage, threads, wheel radius.*

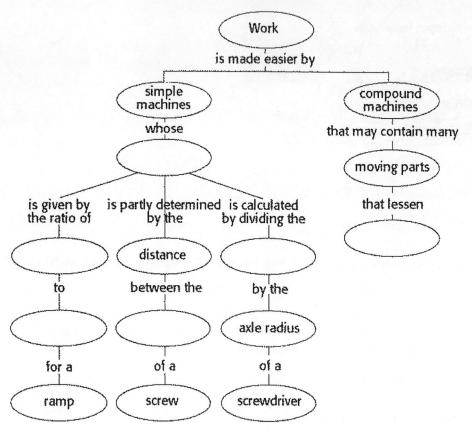

ANS:

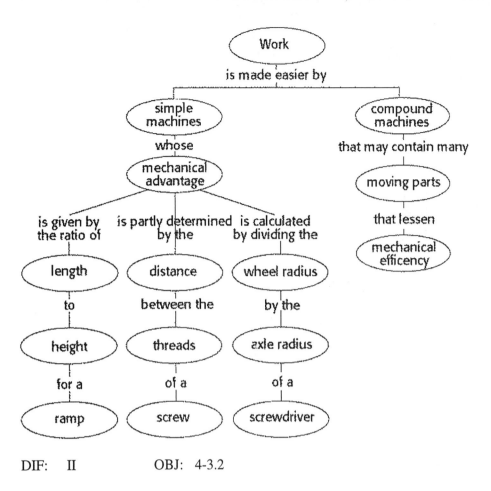

DIF: II   OBJ: 4-3.2

41. Using an ideal pulley system, how much work is done by lifting a load 40 m with a force of 220 N? Show your work.

    ANS:
    Work = force × distance
    220 N × 40 m = 8,800 J

    DIF: II   OBJ: 4-3.2

42. Even though friction is usually considered a drawback in machines, sometimes it is useful. Give an example of a simple machine that relies on friction to function properly. Explain your answer.

    ANS:
    Sample answer: Without friction, cars would not be able to travel up ramps or winding roads, which are both inclined planes. These inclined planes, among others, rely on friction to operate properly.

    DIF: III   OBJ: 4-3.1

43. Examine the pulley arrangement below, and answer the question that follows.

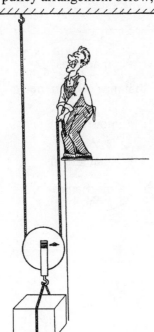

The worker is standing on a platform 4 m above the floor. How many meters of rope must he pull through his hands to raise the block from the floor to his position? Explain.

ANS:
To raise the block 4 m, both sides of the supporting rope have to shorten by 4 m. Therefore, the worker must pull 8 m of rope through his hands.

DIF: II   OBJ: 4-3.2

44. What are the simple machines that make up a pair of scissors?

ANS:
a first-class lever and a wedge

DIF: I   OBJ: 4-3.3

MULTIPLE CHOICE

1. Kinetic energy depends on
   a. mass and volume.
   b. speed and weight.
   c. weight and height.
   d. speed and mass.

   ANS: D   DIF: I   OBJ: 5-1.2

2. Gravitational potential energy depends on
   a. mass and speed.
   b. weight and height.
   c. mass and weight.
   d. height and distance.

   ANS: B   DIF: I   OBJ: 5-1.2

3. Which of the following is NOT a renewable resource?
   a. wind energy
   b. nuclear energy
   c. solar energy
   d. geothermal energy

   ANS: B   DIF: I   OBJ: 5-4.1

4. Which of the following is a conversion from chemical energy to thermal energy?
   a. Food is digested and used to regulate body temperature.
   b. Charcoal is burned in a barbecue pit.
   c. Coal is burned to boil water.
   d. all of the above

   ANS: D   DIF: I   OBJ: 5-2.2

5. Machines can
   a. increase energy.
   b. transfer energy.
   c. convert energy.
   d. Both (b) and (c)

   ANS: D   DIF: I   OBJ: 5-2.3

6. In every energy conversion, some energy is always converted into
   a. kinetic energy.
   b. potential energy.
   c. thermal energy.
   d. mechanical energy.

   ANS: C   DIF: I   OBJ: 5-2.4

7. An object that has kinetic energy must be
   a. at rest.
   b. lifted above the Earth's surface.
   c. in motion.
   d. None of the above

   ANS: C   DIF: I   OBJ: 5-1.2

8. Which of the following is NOT a fossil fuel?
   a. gasoline
   b. coal
   c. firewood
   d. natural gas

   ANS: C   DIF: I   OBJ: 5-4.1

9. Which of the following is NOT an energy resource?
   a. falling water
   b. plant matter
   c. an electric generator
   d. the heat inside the Earth

   ANS: C   DIF: I   OBJ: 5-4.1

10. Suppose you have just lifted a 50 N bowling ball 2 m above the floor. Which of the following is true?
    a. The ball's kinetic energy is equal to 400 J.
    b. Your body is exerting a force of 100 N on the ball.
    c. The ball's gravitational potential energy is equal to 50 J.
    d. You performed work equal to 100 J to lift the ball.

    ANS: D   DIF: I   OBJ: 5-1.1

11. During photosynthesis, plants
    a. combine carbon dioxide with oxygen.
    b. use light energy to produce new substances with chemical energy.
    c. convert thermal energy into kinetic energy.
    d. break down food molecules to release energy.

    ANS: B   DIF: I   OBJ: 5-2.4

12. The kinetic energy of an object can be found if the object's ____ are known.
    a. volume and density
    b. weight and height
    c. speed and mass
    d. distance and time

    ANS: C   DIF: I   OBJ: 5-1.2

13. An object's mechanical energy is
    a. its energy of motion.
    b. not related to the object's mass.
    c. the waste energy it produces by friction.
    d. the sum of its potential and kinetic energies.

    ANS: D   DIF: I   OBJ: 5-1.3

14. Of the following, which type of energy is NOT correctly described?
    a. thermal energy: a measure of particle motion
    b. sound energy: energy that can travel through a vacuum
    c. electrical energy: the energy of moving electrons
    d. light energy: the result of vibrations of electrically charged particles

    ANS: B   DIF: I   OBJ: 5-1.3

15. Suppose you are jumping on a trampoline. At the top of your jump, your
    a. mechanical energy is zero.   c. kinetic and potential energy are equal.
    b. potential energy is at a maximum.   d. potential energy is zero.

    ANS: B   DIF: I   OBJ: 5-2.1

16. When one object does work on another, energy is
    a. destroyed.   c. created.
    b. transferred.   d. All of the above

    ANS: B   DIF: I   OBJ: 5-1.1

17. What is the kinetic energy of a car with a mass of 4000 kg traveling at 20 m/s?
    a. 80,000 J   c. 1,600,000 J
    b. 800,000 J   d. 160,000,000 J

    ANS: B   DIF: II   OBJ: 5-1.2

18. Which of the following vehicles has the MOST kinetic energy?
    a. large truck traveling 30 m/s   c. small car traveling 30 m/s
    b. large truck traveling 25 m/s   d. small car traveling 25 m/s

    ANS: A   DIF: I   OBJ: 5-1.2

19. Which of the following vehicles has the LEAST kinetic energy?
    a. large truck traveling 50 m/s   c. small car traveling 30 m/s
    b. minivan traveling 45 m/s   d. bicycle traveling 10 m/s

    ANS: D   DIF: I   OBJ: 5-1.2

20. Because work and energy are so closely related, they are both expressed in
    a. calories.   c. Calories.
    b. kilocalories.   d. Joules.

    ANS: D   DIF: I   OBJ: 5-1.1

21. ____ is stored in a stretched rubber band.
    a. Kinetic energy   c. Light energy
    b. Potential energy   d. Chemical energy

    ANS: B   DIF: I   OBJ: 5-1.2

22. Before a music box will play music, you must do work on the spring inside by winding it up.
    This work is stored as ____, giving the music box the ability to play music.
    a. potential energy   c. gravitational potential energy
    b. kinetic energy   d. thermal energy

    ANS: A   DIF: I   OBJ: 5-1.2

23. What is the gravitational potential energy of a 540 N diver standing on a platform 3 m high?
    a. 3 J
    b. 180 J
    c. 540 J
    d. 1620 J

    ANS: D   DIF: II   OBJ: 5-1.2

24. Mechanical energy can be
    a. all potential energy.
    b. all kinetic energy.
    c. part potential and part kinetic energy.
    d. All of the above

    ANS: D   DIF: I   OBJ: 5-1.3

25. Which of the following would have the MOST thermal energy?
    a. the particles in an ice cube
    b. the particles in ocean water
    c. the particles in steam
    d. All have an equal amount of thermal energy.

    ANS: B   DIF: I   OBJ: 5-1.3

26. The energy used to cook food in a microwave is a form of
    a. chemical energy.
    b. electrical energy.
    c. sound energy.
    d. light energy.

    ANS: D   DIF: I   OBJ: 5-1.3

27. At what point does a roller coaster have the greatest potential energy?
    a. the top of the biggest hill
    b. the bottom of the biggest hill
    c. the top of the smallest hill
    d. the bottom of the smallest hill

    ANS: A   DIF: I   OBJ: 5-2.2

28. At what point does a roller coaster have the greatest kinetic energy?
    a. the top of the biggest hill
    b. the bottom of the biggest hill
    c. the top of the smallest hill
    d. the bottom of the smallest hill

    ANS: B   DIF: I   OBJ: 5-2.2

29. As a pendulum swings downward, its
    a. potential energy is maximum.
    b. potential energy is converted into kinetic energy.
    c. kinetic energy is maximum.
    d. kinetic energy is converted into potential energy.

    ANS: B   DIF: I   OBJ: 5-2.2

30. The chemical energy of the food we eat is a result of a conversion of
    a. thermal energy to light energy.
    b. light energy to chemical energy.
    c. potential energy to nuclear energy.
    d. potential energy to sound energy.

    ANS: B   DIF: I   OBJ: 5-2.2

31. When rolling downhill, a roller coaster's potential energy is converted into
    a. kinetic energy.
    b. thermal energy.
    c. sound energy.
    d. All of the above

    ANS: D   DIF: I   OBJ: 5-3.1

32. A comparison of the electrical energy going into a light bulb with the light energy coming out of it is a measure of its
    a. friction.
    b. energy resource.
    c. energy efficiency.
    d. perpetual motion.

    ANS: C   DIF: I   OBJ: 5-2.4

33. Our most important energy resource is/are
    a. the sun.
    b. plants.
    c. fossil fuels.
    d. wind.

    ANS: A   DIF: I   OBJ: 5-4.2

34. The cleanest burning fossil fuel is
    a. coal.
    b. oil.
    c. petroleum.
    d. natural gas.

    ANS: D   DIF: I   OBJ: 5-4.3

35. Our most important nonrenewable resource is/are
    a. solar energy.
    b. wind energy.
    c. fossil fuels.
    d. geothermal energy.

    ANS: C   DIF: I   OBJ: 5-4.3

36. The energy source powering the geyser *Old Faithful* in Yellowstone National Park is
    a. electrical energy.
    b. nuclear energy.
    c. wind energy.
    d. geothermal energy.

    ANS: D   DIF: I   OBJ: 5-4.1

37. If fossil fuels are nonrenewable, why do we use them for energy?
    a. They do not produce pollution.
    b. They provide a large amount of thermal energy per unit of mass.
    c. They do not cause acid rain.
    d. all of the above

    ANS: B   DIF: I   OBJ: 5-4.3

38. A single uranium fuel pellet contains the energy equivalent of about one
    a. metric ton of coal.
    b. day's worth of solar energy.
    c. gallon of water.
    d. pound of wood.

    ANS: A   DIF: I   OBJ: 5-4.3

39. Fossil fuels originally received their energy from the
   a. wind.
   c. water.
   b. sun.
   d. Earth's heat.

   ANS: B   DIF: I   OBJ: 5-4.2

40. Plants, wood, and waste are all examples of
   a. nonrenewable resources.
   c. biomass.
   b. inorganic matter.
   d. fossil fuels.

   ANS: C   DIF: I   OBJ: 5-4.1

41. Coal, petroleum, and natural gas are all examples of
   a. fossil fuels.
   b. super-concentrated forms of the sun's energy.
   c. nonrenewable resources.
   d. All of the above

   ANS: D   DIF: I   OBJ: 5-4.1

42. Coal was formed from the remains of ____ that were in swamps millions of years ago.
   a. plants
   c. rocks
   b. animals
   d. dinosaurs

   ANS: A   DIF: I   OBJ: 5-4.1

43. Petroleum was formed from the remains of ____ that were in shallow prehistoric lakes and seas.
   a. silica
   c. organisms
   b. lava flows
   d. metamorphic rock

   ANS: C   DIF: I   OBJ: 5-4.1

44. Most coal used in the United States today is burned by
   a. homes for summertime barbecue.
   b. trains for power.
   c. businesses for power.
   d. power plants to run electric generators.

   ANS: D   DIF: I   OBJ: 5-4.1

45. Natural gas is formed much the same way as ____.
   a. biomass.
   c. nuclear energy.
   b. petroleum.
   d. geothermal energy.

   ANS: B   DIF: I   OBJ: 5-4.1

46. Plastics and synthetic fibers are produced from
   a. coal.
   c. natural gas.
   b. petroleum.
   d. biomass.

   ANS: B   DIF: I   OBJ: 5-4.3

47. Which nonrenewable resource is used the most to heat businesses and homes?
   a. oil
   b. coal
   c. natural gas
   d. petroleum

   ANS: C   DIF: I   OBJ: 5-4.1

48. Which fossil fuel creates the most fuel emissions?
   a. oil
   b. coal
   c. natural gas
   d. petroleum

   ANS: B   DIF: I   OBJ: 5-4.3

49. The United States' primary source of electrical energy is generated by
   a. falling water.
   b. the wind.
   c. burning fossil fuels.
   d. burning biomass.

   ANS: C   DIF: I   OBJ: 5-4.1

50. An electric generator
   a. creates energy.
   b. converts energy of one type into another.
   c. destroys energy during the process.
   d. All of the above

   ANS: B   DIF: I   OBJ: 5-4.3

51. Fossil-fuel and nuclear power plants use _____ to turn a turbine that rotates the generator.
   a. wind
   b. water
   c. steam
   d. electricity

   ANS: C   DIF: I   OBJ: 5-4.3

52. Nonindustrialized countries rely heavily on _____ for energy.
   a. biomass
   b. solar power
   c. nuclear power
   d. geothermal power

   ANS: A   DIF: I   OBJ: 5-4.3

53. Let's say a pin being juggled by a juggler has a potential energy of 10 J and kinetic energy of 400 J. What is the pin's mechanical energy?
   a. 40 J
   b. 390 J
   c. 410 J
   d. 4,000 J

   ANS: C   DIF: II   OBJ: 5-1.3

54. If an object's mechanical energy remains constant and its kinetic energy increases, its
   a. potential energy remains the same.
   b. potential energy increases.
   c. potential energy decreases.
   d. total potential energy plus kinetic energy increases.

   ANS: C   DIF: II   OBJ: 5-1.2

55. If you put a 20 N book on the top shelf of a 2 m high bookshelf, what gravitational potential energy do you give the book?
    a. 10 J
    b. 18 J
    c. 22 J
    d. 40 J

    ANS: D   DIF: II   OBJ: 5-1.3

56. Without the ____ from the sun, life on Earth would not be possible.
    a. nuclear energy
    b. sound energy
    c. electrical energy
    d. wind energy

    ANS: A   DIF: I   OBJ: 5-1.3

57. When you turn on a light bulb, you convert
    a. potential energy into electrical energy and light energy.
    b. electrical energy into kinetic energy and light energy.
    c. electrical energy into light energy and thermal energy.
    d. chemical energy into light energy and thermal energy.

    ANS: C   DIF: I   OBJ: 5-2.2

58. The ____ in the sugars and starches of food fuels all your body functions and movements, and provides the thermal energy that keeps your body temperature constant.
    a. nuclear energy
    b. chemical energy
    c. electrical energy
    d. light energy

    ANS: B   DIF: I   OBJ: 5-1.3

59. Which of the following converts sound energy to electrical energy?
    a. geyser
    b. gas heater
    c. microphone
    d. dam

    ANS: C   DIF: I   OBJ: 5-2.3

60. More efficient light bulbs produce more
    a. thermal energy.
    b. light energy.
    c. waste.
    d. electrical energy.

    ANS: B   DIF: I   OBJ: 5-2.4

61. If you were to climb 3 m up a tree and then pause on a branch to enjoy the view, you would have
    a. kinetic energy
    b. potential energy
    c. light energy
    d. chemical energy

    ANS: B   DIF: I   OBJ: 5-1.2

COMPLETION

1. After a bow is stretched, it has _____. (kinetic energy or potential energy)

    ANS: potential energy   DIF: I   OBJ: 5-2.1

2. Geothermal energy is considered a _____ resource. (renewable or nonrenewable)

    ANS: renewable   DIF: I   OBJ: 5-4.1

3. Coal-burning power plants use steam-driven turbines to generate electrical energy. This is an example of _____. (friction force or energy conversion)

    ANS: energy conversion   DIF: I   OBJ: 5-2.4

4. Petroleum and natural gas are examples of _____. (fossil fuels or mechanical energy)

    ANS: fossil fuels   DIF: I   OBJ: 5-4.1

5. The thermal energy of a substance changes with the _____ of its particles. (gravitational potential energy or kinetic energy)

    ANS: kinetic energy   DIF: I   OBJ: 5-1.3

6. _____ is the ability to do work.

    ANS: Energy   DIF: I   OBJ: 5-1.1

7. _____ energy is the energy of motion.

    ANS: Kinetic   DIF: I   OBJ: 5-1.2

8. _____ energy is the energy an object has because of its position or shape.

    ANS: Potential   DIF: I   OBJ: 5-1.2

9. Calculating an object's _____ is done by calculating the amount of work done on an object in order to lift it to a given height.

    ANS: gravitational potential energy   DIF: I   OBJ: 5-1.2

10. _____ energy is the total energy of motion and position of an object.

    ANS: Mechanical   DIF: II   OBJ: 5-1.2

11. _____ energy is the total kinetic energy of the particles that make up an object.

    ANS: Thermal   DIF: I   OBJ: 5-1.3

12. _____ energy is the energy required to bond particles of matter together and is stored in those bonds.

    ANS: Chemical   DIF: I   OBJ: 5-1.3

13. _____ energy is the energy of moving electrons.

    ANS: Electrical   DIF: I   OBJ: 5-1.3

14. _____ energy is caused by an object's vibrations.

    ANS: Sound   DIF: I   OBJ: 5-1.3

15. _____ energy is produced by the vibrations of electrically charged particles.

    ANS: Light   DIF: I   OBJ: 5-1.3

16. _____ energy is the energy associated with changes in the nucleus of an atom.

    ANS: Nuclear   DIF: I   OBJ: 5-1.3

17. A(n) _____ is a change from one form of energy into another.

    ANS: energy conversion   DIF: I   OBJ: 5-1.3

18. A _____ is a mass hung from a fixed point so that it can swing freely.

    ANS: pendulum   DIF: I   OBJ: 5-2.2

19. _____ uses light energy to produce new substances with chemical energy.

    ANS: Photosynthesis   DIF: I   OBJ: 5-2.4

20. _____ compares the energy before a conversion with the energy after it has been converted.

    ANS: Energy efficiency   DIF: I   OBJ: 5-2.4

21. _____ is a force that opposes the motion between two surfaces that are touching.

    ANS: Friction   DIF: I   OBJ: 5-3.1

22. The _____ states that energy can be neither created nor destroyed.

    ANS: law of conservation of energy   DIF: I   OBJ: 5-3.2

23. If it were possible, a _____ would be a machine that runs forever without any additional energy.

   ANS: perpetual motion machine   DIF: I   OBJ: 5-3.2

24. A(n) _____ is a natural resource that can be converted by humans into other forms of energy in order to do useful work.

   ANS: energy resource   DIF: I   OBJ: 5-4.1

25. Some energy resources, called _____, either cannot be replaced after they are used or can be replaced only over thousands or millions of years.

   ANS: nonrenewable resources   DIF: I   OBJ: 5-4.1

26. _____ are energy resources that formed from the buried remains of plants and animals that lived millions of years ago.

   ANS: Fossil fuels   DIF: I   OBJ: 5-4.1

27. In a process called_____, the nucleus of a uranium atom is split into two smaller nuclei, releasing nuclear energy.

   ANS: nuclear fission   DIF: I   OBJ: 5-4.1

28. _____ can be used and replaced in nature over a relatively short period of time.

   ANS: Renewable resources   DIF: I   OBJ: 5-4.1

29. _____ energy is a result of the heating of the Earth's crust.

   ANS: Geothermal   DIF: I   OBJ: 5-4.1

30. Organic matter that can be burned to release energy is called _____.

   ANS: biomass   DIF: I   OBJ: 5-4.1

31. A _____ is a well-defined group of objects that transfer energy between one another.

   ANS: closed system   DIF: II   OBJ: 5-3.1

32. _____ is electrical energy produced from falling water.

   ANS: Hydroelectricity   DIF: I   OBJ: 5-4.1

Holt Science and Technology
105

SHORT ANSWER

For each pair of terms, explain the differences in their meanings.

1. potential energy/kinetic energy

   ANS:
   Potential energy is energy of position or shape. *Kinetic energy* is energy of motion.

   DIF: I   OBJ: 5-1.2

2. friction/energy conversion

   ANS:
   Friction is a force that opposes motion between two surfaces that are touching. In an *energy conversion*, friction always causes some form of energy to be converted into thermal energy.

   DIF: I   OBJ: 5-2.2

3. energy conversion/law of conservation of energy

   ANS:
   During an *energy conversion*, one form of energy is changed into another form of energy. The *law of conservation of energy* states that energy is neither created nor destroyed during any energy conversion.

   DIF: I   OBJ: 5-3.2

4. energy resources/fossil fuels

   ANS:
   Energy resources are natural resources that can be converted by humans into other forms of energy in order to do useful work. *Fossil fuels* are an energy resource that formed from the remains of organisms that lived millions of years ago.

   DIF: I   OBJ: 5-3.3

5. renewable resources/nonrenewable resources

   ANS:
   Renewable resources are natural resources that can be used and replaced over a relatively short time. *Nonrenewable resources* are natural resources that cannot be replaced or that can be replaced only over thousands or millions of years.

   DIF: I   OBJ: 5-4.1

6. How are energy and work related?

ANS:
Energy is the ability to do work. *Work* is a transfer of energy.

DIF: I   OBJ: 5-1.1

7. What is the difference between kinetic and potential energy?

ANS:
Kinetic energy is the energy of motion, and *potential energy* is an object's energy due to its position or shape.

DIF: I   OBJ: 5-1.2

8. Explain why a high-speed collision might cause more damage to vehicles than a low-speed collision.

ANS:
A high-speed collision would cause more damage to vehicles because the vehicles would have more kinetic energy due to their high speed ($KE = \frac{1}{2}mv^2$). The vehicles would do more work on one another, resulting in large amounts of damage.

DIF: II   OBJ: 5-1.2

9. What determines an object's thermal energy?

ANS:
An object's thermal energy depends on its temperature, the arrangement of its particles, and the number of particles in the object.

DIF: I   OBJ: 5-1.3

10. Describe why chemical energy is a form of potential energy.

ANS:
Sample answer: When a substance forms, work is done to bond particles of matter together. The energy that creates the new bonds is stored in the substance as potential energy.

DIF: I   OBJ: 5-1.3

11. Explain how sound energy is produced when you beat a drum.

ANS:
Sample answer: When you beat a drum, you give it mechanical energy by moving the drumskin back and forth (vibrating). The vibrations cause air particles to vibrate, transmitting energy that results in sound.

DIF: I   OBJ: 5-1.3

12. When you hit a nail into a board using a hammer, the head of the nail gets warm. In terms of kinetic and thermal energy, describe why you think this happens.

ANS:
Sample answer: The kinetic energy of the moving hammer is transferred to the head of the nail, causing particles in the nail to move faster. The faster the particles move, the greater their thermal energy.

DIF: II   OBJ: 5-1.3

13. What is an energy conversion?

ANS:
An energy conversion is a change from one form of energy into another. Any form of energy can be converted into any other form of energy.

DIF: I   OBJ: 5-2.1

14. Describe an example in which electrical energy is converted into thermal energy.

ANS:
Sample answer: In an iron, electrical energy is converted into thermal energy.

DIF: I   OBJ: 5-2.2

15. Describe an energy conversion involving chemical energy.

ANS:
Sample answer: When you light a natural-gas stove, the chemical energy in the natural gas is converted into thermal energy.

DIF: I   OBJ: 5-2.2

16. Describe the kinetic-potential energy conversions that occur when you bounce a basketball.

ANS:
When you bounce a basketball, you give it kinetic energy. At the moment the ball hits the ground, its kinetic energy is greatest and its potential energy is zero. At that moment, the change of shape of the ball that occurs when it hits the ground gives the ball some potential energy that is used as the ball moves back upward. When the ball bounces back up toward your hand, its kinetic energy is converted into potential energy because its position changes. At the moment the ball is at the top of its bounce, its kinetic energy is zero.

DIF: II   OBJ: 5-2.1

17. What is the role of machines in energy conversions?

    ANS:
    Machines can transfer energy from one object to another as they make work easier. For example,
    when you use a crowbar to remove a hubcap, you transfer energy to the crowbar, and the
    crowbar transfers energy to the hubcap. The way the energy is transferred determines how much
    easier the work is to do.

    DIF: I   OBJ: 5-2.3

18. Give an example of a machine that is an energy converter, and explain how the machine converts
    one form of energy to another.

    ANS:
    Accept all reasonable answers. Sample answer: A wind turbine converts the kinetic energy of
    wind into electrical energy.

    DIF: I   OBJ: 5-2.3

19. A car that brakes suddenly comes to a screeching halt. Is the sound energy produced in this
    conversion a useful form of energy? Explain your answer.

    ANS:
    The sound energy of the screeching is not a useful form of energy because it cannot be used to
    do work.

    DIF: II   OBJ: 5-2.4

20. Describe the energy conversions that take place in a pendulum, and explain how the energy is
    conserved.

    ANS:
    Answers will vary, but students should include that potential energy is converted into kinetic
    energy, kinetic energy into potential, and both potential and kinetic into thermal energy because
    of friction. All energy is conserved because some energy becomes thermal energy.

    DIF: I   OBJ: 5-2.1

21. Why is perpetual motion impossible?

    ANS:
    In every system, some of the energy put in is converted into thermal energy that is waste energy.
    A machine cannot run forever unless energy is continually added.

    DIF: I   OBJ: 5-3.1

22. Imagine that you drop a ball. It bounces a few times, but then it stops. Your friend says that the ball has lost all of its energy. Using what you know about the law of conservation of energy, respond to your friend's statement.

ANS:
Answers must include the fact that energy isn't lost in the entire system of the ball, the ground, and the air around the ball; the energy of the ball is converted into other forms.

DIF: II   OBJ: 5-3.1

23. Compare fossil fuels and biomass.

ANS:
Answers will vary but should include the fact that both result from living things and both can be burned to release energy. Fossil fuels are millions of years old, while biomass is organic matter obtained from living things today.

DIF: I   OBJ: 5-4.1

24. Why is nuclear energy a nonrenewable resource?

ANS:
Nuclear energy is a nonrenewable resource because the elements from which it is generated are in limited supply.

DIF: I   OBJ: 5-4.1

25. Trace electrical energy back to the sun.

ANS:
Sample answer: The steam turning the turbine that generated electrical energy comes from water heated by the burning of a fossil fuel, such as coal. The coal is a result of organisms that lived millions of years ago that used light energy from the sun.

DIF: I   OBJ: 5-4.2

26. Use the pie chart below to explain why renewable resources will become more important in years to come.

U.S. Energy Sources

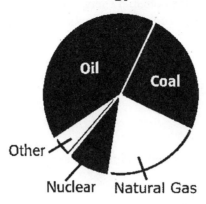

ANS:
When the nonrenewable resources we rely on are used up, we will have to use renewable resources.

DIF: II   OBJ: 5-4.3

27. Compare energy and work. What does one have to do with the other?

ANS:
Energy is the ability to do work. *Work* cannot occur without energy.

DIF: I   OBJ: 5-1.1

28. What is the difference between kinetic and potential energy? Can you describe an activity that utilizes both?

ANS:
Kinetic energy is the energy of motion. *Potential energy* is the energy an object has because of its position or shape. A child's swing shows both kinetic and potential energy.

DIF: I   OBJ: 5-1.2

29. Give an example of an energy conversion that produces a useful result.

ANS:
Answers will vary but students might mention the conversion of chemical energy in their food into the kinetic energy of their movements.

DIF: I   OBJ: 5-2.1

30. Demonstrate the conversion of potential into kinetic energy using a pendulum model.

    ANS:
    As the pendulum is lifted upward, it gains potential energy. When the pendulum is released and swings downward, that potential energy is converted into kinetic energy.

    DIF: I   OBJ: 5-2.1

31. Think of an example other than the ones given in this section to illustrate the law of conservation of energy.

    ANS:
    Answers will vary but should reflect an understanding of energy conservation.

    DIF: I   OBJ: 5-3.2

32. What conditions would have to exist for perpetual motion to be possible?

    ANS:
    For perpetual motion to be possible, there would have to be no waste thermal energy produced and no friction.

    DIF: I   OBJ: 5-3.4

33. Explain the process of fossil fuel formation.

    ANS:
    Organisms that lived millions of years ago died and were covered by layers of sediment. The pressure and the temperatures produced by the overlying layers caused chemical reactions that changed the organic matter into fossil fuel.

    DIF: I   OBJ: 5-4.1

34. Name the five types of energy that are considered renewable resources.

    ANS:
    Renewable resources include solar energy, energy from water, wind energy, geothermal energy, and biomass.

    DIF: I   OBJ: 5-4.1

35. Name two forms of energy, and relate them to kinetic or potential energy.

ANS:
Sample answer: *Thermal energy* depends partly on the kinetic energy of the particles that make up an object. The more *kinetic energy* the particles have, the more thermal energy the object has. *Chemical energy* is a kind of potential energy because when a substance forms, work is done to bond particles of matter together. The energy required to do this work is stored in the new compound as *potential energy* that can be released when the compound is broken down.

DIF: I   OBJ: 5-1.2

36. Give three specific examples of energy conversions.

ANS:
Sample answer: When a person jumps off a diving board, his or her potential energy is converted into kinetic energy. When steam turns the blades of a turbine, the thermal energy of the steam is converted into the kinetic energy of the moving turbine. When a hair dryer is turned on, electrical energy is converted into kinetic energy of the turning fan and thermal energy of the hot coils inside the hair dryer.

DIF: I   OBJ: 5-2.1

37. Explain how energy is conserved within a closed system.

ANS:
A closed system is a well-defined group of objects that transfer energy among one another. Within a closed system, energy is neither created nor destroyed; it just gets converted from one form of energy into another.

DIF: I   OBJ: 5-3.1

38. How are fossil fuels formed?

ANS:
Fossil fuels are the remains of plants and animals that lived millions of years ago. Plants converted energy from the sun by photosynthesis. Animals got this energy from eating plants or other animals. Over millions of years, the remains of these organisms became coal, petroleum, and natural gas.

DIF: I   OBJ: 5-4.1

39. Use the following terms to create a concept map: *energy, machines, energy conversions, thermal energy, friction.*

    ANS:

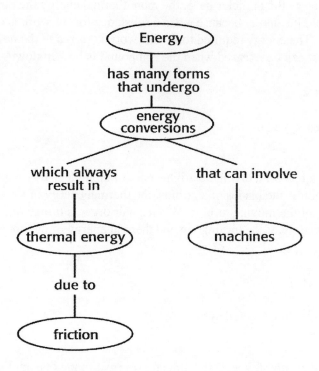

    DIF: I   OBJ: 5-2.4

40. What happens when you blow up a balloon and release it? Describe what you would see in terms of energy.

    ANS:
    Sample answer: When you blow up a balloon, you stretch it. Because you change the balloon's shape, you give it potential energy. When you release the balloon, it zooms around because the potential energy is converted into kinetic energy.

    DIF: II   OBJ: 5-2.1

41. After you coast down a hill on your bike, you eventually come to a complete stop unless you keep pedaling. Relate this to the reason why perpetual motion is impossible.

    ANS:
    Sample answer: Your bike can't keep moving because kinetic energy is converted into thermal energy due to friction between the tires and road.

    DIF: II   OBJ: 5-2.1

42. Look at the picture of the pole-vaulter below. Trace the energy conversions involved in this event, beginning with the pole-vaulter's breakfast of an orange-banana smoothie.

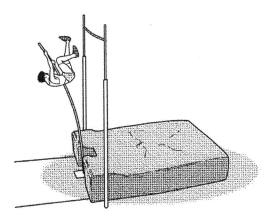

ANS:
Sample answer: The pole vaulter's breakfast provided chemical energy that was converted into kinetic energy as he started his vault. As his pole bends, it stores potential energy. This potential energy is then converted into the kinetic energy that lifts him into the air. As he rises, his kinetic energy is converted into gravitational potential energy. This is converted to kinetic energy as he falls.

DIF: II   OBJ: 5-2.2

43. If the sun were exhausted of its nuclear energy, what would happen to our energy resources on Earth?

ANS:
Sample answer: Without the sun as the source of most energy, we would eventually run out of energy resources.

DIF: II   OBJ: 5-4.2

44. A box has 400 J of gravitational potential energy.
    a. How much work had to be done to give the box that energy?
    b. If the box weighs 100 N, how far was it lifted?

ANS:
a. 400 J
b. 4 m

DIF: II   OBJ: 5-1.1

45. Look at the illustration below, and answer the questions that follow.

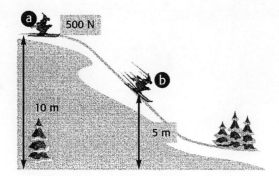

a. What is the skier's gravitational potential energy at point **A**?
b. What is the skier's gravitational potential energy at point **B**?
c. What is the skier's kinetic energy at point **B**?
(Hint: mechanical energy = potential energy + kinetic energy)

ANS:

a. 5,000 J
b. 2,500 J
c. 2,500 J

DIF: II   OBJ: 5-1.2

46. Define and give an example of an energy conversion.

ANS:
An energy conversion is a change of energy from one form into another. For example, in a car engine, the chemical potential energy of gasoline is converted into kinetic energy through combustion. This energy is transferred to the pistons and then to the attached crankshaft, which turns the wheels.

DIF: I   OBJ: 5-2.1

47. How are fossil fuels formed?

ANS:
Energy from the sun is captured and stored by plants, which are sometimes eaten by animals. When these organisms die, their remains are eventually buried. The energy stored in the plant and animal remains is converted to fossil fuels over millions of years.

DIF: I   OBJ: 5-4.1

43. Use the following terms to complete the concept map below: *gravity, kinetic energy, potential energy, a rolling ball, sugar, molecules, chemical bonds.*

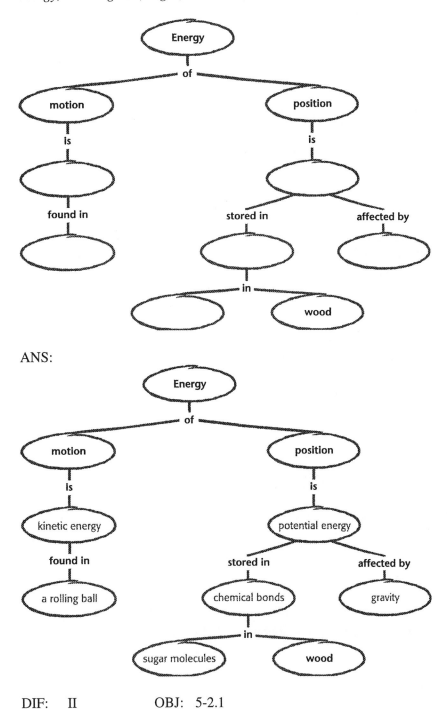

ANS:

DIF: II   OBJ: 5-2.1

Holt Science and Technology
117

DIF: II   OBJ: 5-2.1

49. How much work is done on a 30 N rock that is lifted 1.5 m off the ground? Show your work.

ANS:
Work = force × distance = 30 N × 1.5 m = 45 J of work done on the rock

DIF: II   OBJ: 5-1.1

50. A weightlifter does work to lift a barbell over his head. Suppose the weightlifter grows several centimeters taller.
   a. How would this affect the work he does to lift the same barbell? Explain.
   b. How would this affect the potential energy of the raised barbell? Explain.

ANS:

   a. The force needed to lift the weight of the barbell would not change, but the distance the barbell is lifted would change. This change in distance would increase the amount of work performed by the weightlifter.
   b. The gravitational potential energy of the barbell would increase as well.

DIF: I   OBJ: 5-1.2

Use the diagram below to answer the following questions.

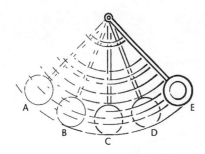

51. At which point(s) does the pendulum have the most potential energy?

ANS:
The pendulum has maximum potential energy at points **A** and **E**.

DIF: I   OBJ: 5-2.1

52. At which point(s) does the pendulum have the most kinetic energy?

ANS:
The pendulum has maximum kinetic energy at point **C**.

DIF: II   OBJ: 5-2.1

TRUE/FALSE

1. Heat is the transfer of energy between two objects at different temperatures.

    ANS: T   DIF: I   OBJ: 6-2.1

2. Radiation occurs in fluids.

    ANS: F   DIF: I   OBJ: 6-2.2

3. Convection currents result from temperature differences in liquids and gases.

    ANS: T   DIF: I   OBJ: 6-2.2

4. Radiation is the means by which the energy from the sun is transferred to Earth.

    ANS: T   DIF: I   OBJ: 6-2.2

5. Water stays warm or cool longer than land does, because water has a lower specific heat capacity than land does.

    ANS: F   DIF: I   OBJ: 6-2.3

6. When ice melts, it absorbs energy.

    ANS: T   DIF: I   OBJ: 6-3.2

7. When a liquid evaporates, it absorbs energy.

    ANS: T   DIF: I   OBJ: 6-3.2

8. When a vapor condenses to a liquid, energy is given off.

    ANS: T   DIF: I   OBJ: 6-3.2

9. When a liquid boils, energy is absorbed.

    ANS: T   DIF: I   OBJ: 6-3.2

10. You can cool the kitchen by leaving the refrigerator door open.

    ANS: F   DIF: I   OBJ: 6-4.3

11. Refrigeration is possible because of energy absorbed and released during changes in state.

    ANS: T   DIF: I   OBJ: 6-4.3

12. A radiator heats a room by heating the air, which circulates in convection currents.

ANS: T   DIF: I   OBJ: 6-4.1

MULTIPLE CHOICE

1. Which of the following temperatures is the lowest?
   a. 100°C   c. 100 K
   b. 100°F   d. They are the same.

ANS: C   DIF: I   OBJ: 6-1.3

2. Compared with the Pacific Ocean, a cup of hot chocolate has
   a. more thermal energy and a higher temperature.
   b. less thermal energy and a higher temperature.
   c. more thermal energy and a lower temperature.
   d. less thermal energy and a lower temperature.

ANS: B   DIF: I   OBJ: 6-2.1

3. The energy units on a food label are
   a. degrees.   c. calories.
   b. Calories.   d. joules.

ANS: B   DIF: I   OBJ: 6-2.4

4. Which of the following materials would NOT be a good insulator?
   a. wood   c. metal
   b. cloth   d. rubber

ANS: C   DIF: I   OBJ: 6-2.2

5. The engine in a car is a(n)
   a. heat engine.   c. internal combustion engine.
   b. external combustion engine.   d. Both (a) and (c)

ANS: D   DIF: I   OBJ: 6-3.2

6. Materials that warm up or cool down very quickly have a
   a. low specific heat capacity.   c. low temperature.
   b. high specific heat capacity.   d. high temperature.

ANS: A   DIF: I   OBJ: 6-2.3

7. In an air conditioner, thermal energy is
   a. transferred from higher to lower temperatures.
   b. transferred from lower to higher temperatures.
   c. used to do work.
   d. taken from air outside a building and transferred to air inside the building.

ANS: B   DIF: I   OBJ: 6-4.3

8. A tile floor feels colder on your bare feet than a wooden floor because the tile floor
   a. is a better insulator.
   b. has a smoother surface than a wooden floor.
   c. conducts more energy from your feet.
   d. will always have a lower temperature than a wooden floor.

   ANS: C   DIF: I   OBJ: 6-2.2

9. The lowest possible temperature is
   a. 273 K.   c. 0°C.
   b. 0°F.   d. 0 K.

   ANS: D   DIF: I   OBJ: 6-1.3

10. In which state does water have the lowest average kinetic energy?
   a. liquid   c. solid
   b. gas   d. Each state has the same kinetic energy.

   ANS: C   DIF: I   OBJ: 6-3.1

11. When water vapor condenses,
   a. the surrounding air is cooled.   c. the surrounding air is warmed.
   b. the water molecules spread apart. d. a solid is formed.

   ANS: C   DIF: I   OBJ: 6-3.2

12. Which of the following represents the greatest amount of energy?
   a. 1,000 cal   c. 100 kcal
   b. 500 Cal   d. 100,000 J

   ANS: B   DIF: I   OBJ: 6-2.2

13. Which of the following processes is NOT a physical change?
   a. melting   c. freezing
   b. boiling   d. burning

   ANS: D   DIF: I   OBJ: 6-3.2

14. If two substances have the same temperature, then
   a. the substances have the same average kinetic energy.
   b. their particles will not react chemically.
   c. the substances have equal thermal energies.
   d. their specific heats are identical.

   ANS: A   DIF: I   OBJ: 6-2.3

15. The official SI temperature scale is the
   a. Celsius scale.   c. Faraday scale.
   b. Fahrenheit scale.   d. Kelvin scale.

   ANS: D   DIF: I   OBJ: 6-1.1

16. Alcohol in a thermometer rises as a result of
    a. thermal contraction.
    b. thermal expansion.
    c. absolute contraction.
    d. absolute expansion.

    ANS: B   DIF: I   OBJ: 6-1.2

17. Absolute zero is
    a. 0 K.
    b. 0°C.
    c. 0°F.
    d. All of the above

    ANS: A   DIF: I   OBJ: 6-1.2

18. Which of the following equations can be used to calculate the energy transferred by heat?

    a. Energy transferred = specific heat capacity × mass × change in temperature

    b. Energy transferred = $\dfrac{\text{specific heat capacity}}{\text{mass} \times \text{change in temperature}}$

    c. Energy transferred = $\dfrac{\text{mass} \times \text{change in temperature}}{\text{specific heat capacity}}$

    d. Energy transferred = $\dfrac{\text{change in temperature}}{\text{mass} \times \text{specific heat capacity}}$

    ANS: A   DIF: I   OBJ: 6-2.3

19. Which of the following is used to find the specific heat capacity of a substance?
    a. balance
    b. anemometer
    c. calorimeter
    d. barometer

    ANS: C   DIF: I   OBJ: 6-2.4

20. Some bridges are constructed with small gaps called expansion joints because
    a. it prevents the bridge from falling during an earthquake.
    b. concrete segments expand on hot days.
    c. it allows the bridge to raise and lower for passing boats.
    d. it makes a "thuh-thunk" noise as you drive over it.

    ANS: B   DIF: I   OBJ: 6-1.2

21. A bimetallic strip in a thermostat uncoils because of
    a. thermal expansion.
    b. thermal contraction.
    c. absolute expansion.
    d. absolute contraction.

    ANS: A   DIF: I   OBJ: 6-1.2

22. Convert 68°F to °C.
    a. 180°C
    c. 56°C
    b. 70°C
    d. 20°C

    ANS: D   DIF: I   OBJ: 6-1.3

23. Because particles are in motion, they have
    a. thermal expansion.
    c. kinetic energy.
    b. thermal contraction.
    d. potential energy.

    ANS: C   DIF: I   OBJ: 6-1.1

24. The temperature of a substance is determined by
    a. how much of the substance you have.
    b. whether the substance is a solid, liquid, or a gas.
    c. how fast the atoms or molecules of the substance are moving.
    d. the potential energy of the substance.

    ANS: C   DIF: I   OBJ: 6-1.1

25. The motion of particles in a substance is
    a. random.
    c. perpendicular.
    b. the same.
    d. parallel.

    ANS: A   DIF: I   OBJ: 6-1.1

26. Thermostats use two metals in a bimetallic strip because
    a. different materials expand at different rates, causing the strip to uncoil.
    b. it looks better.
    c. one metal measures the temperature, while the other metal closes the circuit.
    d. None of the above

    ANS: A   DIF: I   OBJ: 6-1.2

27. Heat is
    a. the same thing as temperature.
    b. the same as thermal expansion.
    c. the transfer of energy between objects that are at different temperatures.
    d. All of the above

    ANS: C   DIF: I   OBJ: 6-2.1

28. The total kinetic energy of the particles that make up a substance is
    a. thermal energy.
    c. temperature.
    b. thermal expansion.
    d. heat.

    ANS: A   DIF: I   OBJ: 6-2.4

29. Which of the following has the most thermal energy?
    a. a pot of soup   c. a plateful of soup
    b. a bowl of soup   d. a spoonful of soup

    ANS: A   DIF: I   OBJ: 6-2.4

30. You add ice to the drinks in an insulated cooler because you hope that
    a. the ice and drinks will reach thermal equilibrium.
    b. the drinks will transfer heat to the ice, thereby cooling the drinks.
    c. the kinetic energy of the drink's molecules will be reduced.
    d. All of the above

    ANS: D   DIF: I   OBJ: 6-2.1

31. At thermal equilibrium,
    a. the thermal energy of the warmer object increases.
    b. the thermal energy of the warmer object decreases.
    c. no net change in either object's thermal energy occurs.
    d. None of the above

    ANS: C   DIF: I   OBJ: 6-2.1

32. When two objects reach thermal equilibrium,
    a. thermal energy is no longer transferred between the two objects.
    b. both objects have the same temperature.
    c. one object may have more thermal energy.
    d. All of the above

    ANS: D   DIF: I   OBJ: 6-2.1

33. The transfer of thermal energy from one substance to another through direct contact is
    a. radiation.   c. convection.
    b. conduction.   d. insulation.

    ANS: B   DIF: I   OBJ: 6-2.2

34. The transfer of thermal energy by the movement of a liquid or a gas is
    a. convection.   c. radiation.
    b. insulation.   d. conduction.

    ANS: A   DIF: I   OBJ: 6-2.2

35. The transfer of thermal energy through space is
    a. convection.   c. radiation.
    b. conduction.   d. insulation.

    ANS: C   DIF: I   OBJ: 6-2.2

36. The handle of a metal spoon warms up when it is placed in a hot bowl of soup because of
    a. convection.   c. insulation.
    b. conduction.   d. radiation.

    ANS: B   DIF: I   OBJ: 6-2.2

37. One possible cause of tectonic plate motion is the rise of hot material from deep within the Earth while cooler material near the surface sinks. As this process repeats, it is an example of
    a. convection.   c. radiation.
    b. conduction.   d. insulation.

    ANS: A   DIF: I   OBJ: 6-2.2

38. Farmers grow vegetables in winter using greenhouses, which receive heat by
    a. convection.   c. radiation.
    b. conduction.   d. insulation.

    ANS: C   DIF: I   OBJ: 6-2.2

39. Warming up beside a campfire is an example of
    a. insulation.   c. radiation.
    b. conduction.   d. convection.

    ANS: C   DIF: I   OBJ: 6-2.2

40. Which of the following is NOT a conductor?
    a. curling iron   c. copper pipe
    b. iron skillet   d. oven mitt

    ANS: D   DIF: I   OBJ: 6-2.2

41. Water has a specific heat capacity of 4,184 J/kg•°C. This means that it requires
    a. 1 J of energy to raise 1 kg of water by 4,184°C.
    b. 1 J of energy to raise 4,184 kg of water by 1°C.
    c. 4,184 J of energy to raise 1 kg of water by 1°C.
    d. 4,184 J of energy to raise 1 kg of water by 1°F.

    ANS: C   DIF: I   OBJ: 6-2.3

42. Particles of a _____ do not move fast enough to overcome the strong attraction between them, so they are held tightly together.
    a. liquid   c. gas
    b. solid   d. plasma

    ANS: B   DIF: I   OBJ: 6-3.1

43. Particles of a _____ vibrate in place.
    a. solid   c. gas
    b. liquid   d. plasma

    ANS: A   DIF: I   OBJ: 6-3.1

44. Particles of a ____ move fast enough to overcome some of the attraction between them.
    a. gas
    b. liquid
    c. solid
    d. plasma

    ANS: B   DIF: I   OBJ: 6-3.1

45. Particles of a ____ are able to slide past one another.
    a. plasma
    b. liquid
    c. gas
    d. solid

    ANS: B   DIF: I   OBJ: 6-3.1

46. Particles of a ____ move fast enough to overcome nearly all of the attraction between them.
    a. solid
    b. liquid
    c. gas
    d. Both (a) and (b)

    ANS: C   DIF: I   OBJ: 6-3.1

47. Particles of a ____ move independently of one another.
    a. solid
    b. liquid
    c. gas
    d. All of the above

    ANS: C   DIF: I   OBJ: 6-3.1

48. Particles move slowest in a
    a. solid.
    b. liquid.
    c. gas.
    d. plasma.

    ANS: A   DIF: I   OBJ: 6-3.1

49. A change of state is a
    a. physical change.
    b. chemical change.
    c. process by which two states of matter coexist.
    d. process that converts one substance into an entirely new substance that cannot be changed back again.

    ANS: A   DIF: I   OBJ: 6-3.2

50. Ice begins to melt at 0°C. When an ice cube has just fully melted, its temperature
    a. is the same as the melting point.
    b. is greater than the melting point.
    c. is less than the melting point.
    d. fluctuates.

    ANS: A   DIF: I   OBJ: 6-3.2

51. Liquid butter becomes solid when it is in a refrigerator. Therefore, the freezing point of butter
    a. is at the same temperature as the freezing point of ice.
    b. is at a higher temperature than the freezing point of ice.
    c. is at a lower temperature than the freezing point of ice.
    d. cannot be compared with ice.

    ANS: B   DIF: II   OBJ: 6-3.2

52. You can get a more severe burn from steam than from boiling water because
    a. steam contains more energy per unit mass than boiling water.
    b. steam contains less energy per unit mass than boiling water.
    c. steam contains the same amount of energy per unit mass than boiling water.
    d. None of the above

    ANS: A   DIF: I   OBJ: 6-3.2

53. Water is typically used in heating systems because it has a ____ specific heat capacity when
    compared with other substances.
    a. low   c. equal
    b. moderate   d. high

    ANS: D   DIF: I   OBJ: 6-4.1

54. In a warm-air heating system, warm air is circulated through the rooms by ____ currents.
    a. conduction   c. insulation
    b. convection   d. radiation

    ANS: B   DIF: I   OBJ: 6-4.1

55. ____ systems do NOT have moving parts.
    a. Passive solar heating   c. Warm-air heating
    b. Active solar heating   d. Hot-water heating

    ANS: A   DIF: I   OBJ: 6-4.1

56. One type of engine used to generate electrical energy at a power plant is a
    a. modern form of a steam engine.   c. internal combustion engine.
    b. external combustion engine.   d. Both (a) and (b)

    ANS: D   DIF: I   OBJ: 6-4.2

57. Coastal areas stay cool during the summer when inland temperatures soar. These areas stay
    moderately warm in winter, even when inland temperatures drop. This happens because
    a. land has a higher specific heat than water.
    b. water has a higher specific heat than land.
    c. there is more water than land.
    d. there is more land than water.

    ANS: B   DIF: I   OBJ: 6-2.3

58. Much of the hot water in Reykjavík, Iceland comes from wells bored into the hot springs of Reykir. The water temperature from the wells is 188.6°F. Express this temperature in degrees Celsius.
   a. −84.4°C
   b. 87°C
   c. 371.48°C
   d. 461.6°C

   ANS: B   DIF: II   OBJ: 6-1.3

59. The Hawaiian lavas at Kilauea Crater have the highest temperatures measured on Earth's surface—over 2192°F! Express this temperature in degrees Celsius.
   a. 1200°C
   b. 1919°C
   c. 2465°C
   d. 3977.6°C

   ANS: A   DIF: II   OBJ: 6-1.3

60. *Hypothermia* is a condition in which a person's body temperature drops below 95°F. If a person's body temperature drops below 87.8°F, they can die. What is this fatal temperature in Kelvin?
   a. −242 K
   b. −185.2 K
   c. 360.8 K
   d. 304 K

   ANS: D   DIF: II   OBJ: 6-1.3

61. Just as the human body cannot survive if its temperature falls to too low a temperature, it also cannot survive if its temperature is too high. In a condition called *hyperthermia*, energy is transferred to the body from its surroundings, causing the body's temperature to increase. The condition known as heat stroke is a severe form of hyperthermia. Normally, the human body cannot survive for long at a temperature of about 42°C, although a recent survivor of heat stroke had a high temperature of nearly 47°C. Express the survivor's temperature in Kelvin.
   a. −231 K
   b. −226 K
   c. 315 K
   d. 320 K

   ANS: D   DIF: II   OBJ: 6-1.3

62. On July 10, 1913, the temperature reached 330 K in Death Valley, California—the hottest temperature ever reached in the United States. Calculate this temperature in degrees Celsius.
   a. 57°C
   b. 165.6°C
   c. 603°C
   d. 626°C

   ANS: A   DIF: II   OBJ: 6-1.3

63. Air that slowly falls from high altitudes can result in cold fronts that sweep across many states. At 30,000 feet above the Earth's surface, the air temperature can be around −67°F. Find this temperature in Celsius.
   a. −55°C
   b. −69°C
   c. −5°C
   d. 206°C

   ANS: A   DIF: II   OBJ: 6-1.3

64. In 1906, a 0.6 kg diamond was found at the Premier Mine in South Africa—the world's largest uncut diamond. Suppose the diamond's temperature changed by 2°C as it was cut. If the heat capacity of diamond is 710 J/kg•°C, how much energy was transferred to the diamond by heat?
   a. 1.2 J
   b. 712.6 J
   c. 852 J
   d. 1420 J

   ANS: C   DIF: II   OBJ: 6-2.3

65. A 0.3 kg piece of copper is heated and fashioned into a bracelet. The amount of energy transferred by heat to the copper is 66,300 J. If the specific heat of copper is 390 J/kg•°C, what is the change of the copper's temperature?
   a. 51°C
   b. 170°C
   c. 567°C
   d. 1300°C

   ANS: C   DIF: III   OBJ: 6-2.3

66. The water in a swimming pool transfers 1.1×10^{10} J of energy by heat to the cool night air. If the temperature of the water, which has a specific heat of 4186 J/kg•°C, decreases by 10°C, what is the mass of the water in the pool?
   a. 418.6 kg
   b. 46,046 kg
   c. 41,860 kg
   d. 262,780 kg

   ANS: D   DIF: III   OBJ: 6-2.3

67. Two substances can have the same temperature but different amounts of thermal energy because temperature, unlike thermal energy,
   a. depends on weight.
   b. depends on density.
   c. depends on mass.
   d. does not depend on mass.

   ANS: D   DIF: I   OBJ: 6-2.4

68. A small amount of a substance at a particular temperature will have _____ a large amount of the substance at the same temperature.
   a. less thermal energy than
   b. more thermal energy than
   c. the same thermal energy as
   d. no thermal energy, unlike

   ANS: A   DIF: I   OBJ: 6-2.4

69. Which of the following has the LEAST thermal energy?
   a. a drop of boiling water
   b. a cup of boiling water
   c. a pot of boiling water
   d. all have the same thermal energy

   ANS: A   DIF: I   OBJ: 6-2.4

70. Which of the following does NOT vary with the mass of a substance?
   a. heat
   b. energy
   c. thermal energy
   d. temperature

   ANS: D   DIF: I   OBJ: 6-2.4

71. The temperature of the air in a city may be higher than in the surrounding countryside because
    a. of the heat island effect.
    b. excessive amounts of waste thermal energy are added to the urban environment.
    c. of automobiles, factories, home heating and cooling, and the local population.
    d. All of the above

    ANS: D   DIF: I   OBJ: 6-4.4

72. If you had a refrigerator in Antarctica, you would have to ____ to keep it running.
    a. keep it outdoors   c. surround it with conductive materials
    b. heat it   d. use ice

    ANS: B   DIF: II   OBJ: 6-4.3

73. The area near the back of a refrigerator feels warm because
    a. of the temperature change from the cool refrigerator and the warm air around the box.
    b. it transfers thermal energy from outside the refrigerator to the condenser coils.
    c. it transfers thermal energy from inside the refrigerator to the condenser coils.
    d. liquid becomes cold as it is compressed by a compressor.

    ANS: C   DIF: I   OBJ: 6-4.3

74. In a ____, fuel combines with oxygen in a chemical change that produces thermal energy.
    a. heat engine   c. active solar heating system
    b. refrigerator   d. passive solar heating system

    ANS: A   DIF: I   OBJ: 6-4.1

COMPLETION

1. Most substances _____ when they are cooled. (expand or contract)

    ANS: contract   DIF: I   OBJ: 6-1.2

2. The common temperature scale used by most Americans is the _____ scale. (Fahrenheit or Celsius or Kelvin)

    ANS: Fahrenheit   DIF: I   OBJ: 6-1.3

3. Scientists use either the _____ scale or the _____ scale. (Fahrenheit or Celsius or Kelvin)

    ANS: Celsius, Kelvin   DIF: I   OBJ: 6-1.3

4. Temperature _____ as average kinetic energy decreases. (increases or decreases)

    ANS: decreases   DIF: I   OBJ: 6-1.1

5. The condensation of water is a _____ _____. (state of matter or change of state)

   ANS: change of state   DIF: I   OBJ: 6-3.2

6. Aluminum changes temperature more slowly than lead; therefore, aluminum has a higher _____ than lead. (thermal energy or specific heat capacity)

   ANS: specific heat capacity   DIF: I   OBJ: 6-2.3

7. When one end of an iron nail is held in a flame, the energy is transmitted along the nail by _____. (convection or conduction)

   ANS: conduction   DIF: I   OBJ: 6-2.2

8. Thermal energy flows between objects that differ in _____. (thermal expansion or temperature)

   ANS: temperature   DIF: I   OBJ: 6-2.1

9. A(n) _____ converts thermal energy into mechanical work. (insulator coil or heat engine)

   ANS: heat engine   DIF: I   OBJ: 6-4.2

10. _____ is a measure of the average kinetic energy of the particles in an object.

    ANS: Temperature   DIF: I   OBJ: 6-1.1

11. The increase in volume of a substance due to an increase in temperature is called _____.

    ANS: thermal expansion   DIF: I   OBJ: 6-1.2

12. The amount of energy needed to change the temperature of 1 kg of a substance by 1°C is its _____.

    ANS: specific heat capacity   DIF: I   OBJ: 6-2.3

13. The _____ of matter are the physical forms in which a substance can exist.

    ANS: states   DIF: I   OBJ: 6-3.1

14. A(n) _____ is the conversion of a substance from one physical form to another.

    ANS: change of state   DIF: I   OBJ: 6-3.2

15. A(n) _____ is a substance that reduces the transfer of thermal energy.

    ANS: insulator   DIF: I   OBJ: 6-4.1

16. _____ combustion engines burn fuel outside the engine.

    ANS: External   DIF: I   OBJ: 6-4.2

17. A negative effect of excess thermal energy from heat technology on the environment is
    _____.

    ANS: thermal pollution   DIF: I   OBJ: 6-4.4

18. Energy transferred by _____ varies with the mass, specific heat capacity, and
    temperature change of a substance.

    ANS: heat   DIF: I   OBJ: 6-2.4

19. _____ varies with the mass and temperature of a substance.

    ANS: Thermal energy   DIF: I   OBJ: 6-2.4

20. A(n) _____ is a gas that has a boiling point below room temperature and is
    used in cooling systems.

    ANS: refrigerant   DIF: I   OBJ: 6-4.3

21. Engines burn fuel through a process called _____.

    ANS: combustion   DIF: I   OBJ: 6-4.2

22. _____ solar heating systems do not have moving parts. Instead, these systems
    utilize thick walls and large windows that face south.

    ANS: Passive   DIF: I   OBJ: 6-4.1

23. _____ solar heating systems often consist of solar collectors, a network of
    pipes, a fan, and a water storage tank.

    ANS: Active   DIF: I   OBJ: 6-4.1

SHORT ANSWER

For each pair of terms, explain the differences in their meanings.

1. temperature/thermal energy

   ANS:
   Temperature is a direct measure of the average kinetic energy of the particles of a substance. *Thermal energy* is the total kinetic energy of the particles of the substance.

   DIF: I   OBJ: 6-1.1

2. heat/thermal energy

   ANS:
   Heat is the transfer of energy between objects at different temperatures. *Thermal energy* is energy transferred by heat.

   DIF: I   OBJ: 6-2.1

3. conductor/insulator

   ANS:
   A *conductor* is a material that conducts energy easily. An *insulator* is a material that does not conduct energy easily.

   DIF: I   OBJ: 6-2.2

4. conduction/convection

   ANS:
   Conduction is the transfer of energy from one substance to another through direct contact. *Convection* is the transfer of energy by the movement of a gas or a liquid.

   DIF: I   OBJ: 6-2.2

5. states of matter/change of state

   ANS:
   The *states of matter* are the physical forms in which a substance can exist. A *change of state* occurs when a substance changes from one state to another.

   DIF: I   OBJ: 6-3.2

6. What is temperature?

    ANS:
    Temperature is a measure of how hot or cold an object is. Specifically, temperature is a direct measure of the average kinetic energy of the particles in an object.

    DIF: I   OBJ: 6-1.1

7. What is the coldest temperature possible?

    ANS:
    The coldest temperature possible is absolute zero (0 K or –273°C).

    DIF: I   OBJ: 6-1.3

8. Convert 35°C to degrees Fahrenheit.

    ANS:
    °F = (9/5 × °C) + 32
    °F = 9/5 × 35°C + 32
    °F = 95, or 95°F

    DIF: I   OBJ: 6-1.3

9. Why do you think heating a full pot of soup on the stove could cause the soup to overflow?

    ANS:
    The soup could overflow its pot as it cooks on the stove because of thermal expansion. The soup will expand in volume as its temperature increases. If the cold soup is too close to the top of a pot, it will likely overflow as it expands.

    DIF: II   OBJ: 6-1.2

10. What is heat?

    ANS:
    Heat is the transfer of energy between objects that are at different temperatures. Energy is always transferred from an object with a higher temperature to an object with a lower temperature.

    DIF: I   OBJ: 6-2.1

11. Explain how radiation is different from conduction and convection.

    ANS:
    Radiation is different from conduction and convection in that it does not require an energy transfer among particles of matter. Radiation is the transfer of energy through matter or space as electromagnetic waves, such as visible light or infrared waves.

    DIF: I   OBJ: 6-2.2

12. Why do many metal cooking utensils have wooden handles?

ANS:
Sample answer: Many metal cooking utensils have wooden handles because wood is an insulator. When you are preparing hot food, a wooden handle will prevent the thermal energy of the food from being conducted to your hand.

DIF: II   OBJ: 6-2.2

13. Some objects get hot more quickly than others. Why?

ANS:
It depends on an object's thermal conductivity. An object with a low thermal conductivity gets hotter (and cooler) more slowly than one with a high thermal conductivity.

DIF: I   OBJ: 6-2.3

14. How are temperature and heat different?

ANS:
Temperature is a measure of the average kinetic energy of particles in a substance. Heat is the transfer of energy between objects that are at different temperatures (or the amount of energy transferred) and must be calculated, not measured.

DIF: I   OBJ: 6-2.4

15. How do you think the specific heat capacities for water and air influence the temperature of a swimming pool and the area around it?

ANS:
Sample answer: Water has a higher specific heat capacity than air. The water must absorb more energy to increase its temperature than does the surrounding air, so water may feel cool even when it is hot outdoors.

DIF: II   OBJ: 6-2.3

16. During a change of state, why doesn't the temperature of the substance change?

ANS:
Energy added to or removed from a substance during a change of state rearranges the particles of the substance rather than raising or lowering the temperature.

DIF: I   OBJ: 6-3.2

17. Compare the thermal energy of 10 g of ice with the thermal energy of the same amount of water.

ANS:
Particles of water in the liquid state have more kinetic energy than particles of water in the solid state, so 10 mL of water has more thermal energy than 10 g of ice (if the water and ice are not at the same temperature.

DIF: I   OBJ: 6-2.5

18. When water evaporates (changes from a liquid to a gas), the air near the water's surface becomes cooler. Explain why this happens.

ANS:
Sample answer: For water to evaporate, it must absorb energy. The air near the water's surface transfers energy to the water to make it evaporate. Because the air loses energy, it becomes cooler.

DIF: I   OBJ: 6-3.2

19. Many cold packs used for sports injuries are activated by bending the package, causing the substances inside to interact. How is heat involved in this process?

ANS:
Sample answer: When you bend the ice pack, the substances inside interact. That interaction absorbs so much energy that the pack feels colder.

DIF: II   OBJ: 6-3.3

20. Compare a hot-water heating system with a warm-air heating system.

ANS:
Accept all reasonable responses.

DIF: I   OBJ: 6-4.1

21. What is the difference between an external combustion engine and an internal combustion engine?

ANS:
In an *external combustion engine,* fuel is burned outside the engine. In an *internal combustion engine,* fuel is burned inside the engine.

DIF: I   OBJ: 6-4.2

22. How are changes of state an important part of the way a refrigerator works?

ANS:
Sample answer: When the compressed liquid refrigerant passes through the expansion valve, it changes into a gas. During this change of state, the refrigerant absorbs thermal energy from the interior of the refrigerator, keeping the interior cold.

DIF: II   OBJ: 6-4.3

23. How does temperature relate to kinetic energy?

ANS:
Temperature is a direct measure of the average kinetic energy of the particles in a substance. The more kinetic energy the particles have, the higher the temperature of the substance.

DIF: I   OBJ: 6-1.1

24. What is specific heat capacity?

ANS:
Specific heat capacity is the amount of energy needed to change the temperature of 1 kg of a substance by 1°C. Specific heat capacity determines the rate at which a substance changes temperature. Every substance has a unique specific heat capacity.

DIF: I   OBJ: 6-2.3

25. Explain how heat affects matter during a change of state.

ANS:
During a change of state, the thermal energy transferred to or from the matter is used to rearrange the particles of the matter. This rearranging involves overcoming the attraction between particles (as when a solid changes to a liquid) or increasing the attraction between particles (as when a liquid changes to a solid).

DIF: I   OBJ: 6-3.2

26. Describe how a bimetallic strip works in a thermostat.

ANS:
A bimetallic strip is made of two metals that expand and contract at different rates with changes in temperature. If the temperature drops below the thermostat setting, the strip coils up. This causes a glass tube to tilt, and a drop of mercury rolls down the tube to close an electric circuit that turns on the heater. When the temperature rises, the process is reversed.

DIF: II   OBJ: 6-1.3

27. On a separate sheet of paper, create a concept map using the following terms: *thermal energy, temperature, radiation, heat, conduction, convection.*

ANS:

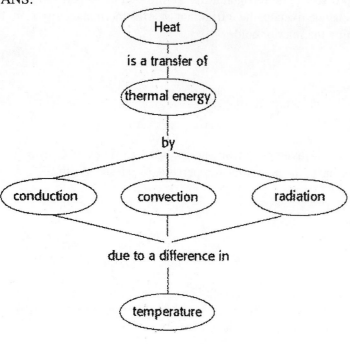

DIF: II   OBJ: 6-2.2

28. Why does placing a jar under warm running water help to loosen the lid on the jar?

ANS:
The water warms the lid, causing it to expand so that it can be removed more easily.

DIF: II   OBJ: 6-1.3

29. Why do you think a down-filled jacket keeps you so warm?
(Hint: Think about what insulation does.)

ANS:
Inside a down-filled jacket are thousands of air pockets between the feathers. You stay warm because these air pockets slow the transfer of energy from your body to the cooler air outside the jacket.

DIF: II   OBJ: 6-2.2

Holt Science and Technology
138

30. Would opening the refrigerator cool a room in a house? Why or why not?

    ANS:
    No; a refrigerator transfers thermal energy from its interior to the room. If you open the refrigerator door, its interior warms up, and it has to transfer even more thermal energy away from its interior.

    DIF: II   OBJ: 6-4.3

31. In a hot-air balloon, air is heated by a flame. Explain how this enables the balloon to float in the air.

    ANS:
    Heating the air increases the kinetic energy of the air particles, causing them to move faster and spread apart. As a result, the warmer air rises and the balloon floats.

    DIF: II   OBJ: 6-1.2

32. The weather forecast calls for a temperature of 86°F. What is the corresponding temperature in degrees Celsius? in kelvins?

    ANS:
    30°C; 303 K

    DIF: II   OBJ: 6-1.3

33. Suppose 1,300 mL of water are heated from 20°C to 100°C. How much energy was transferred to the water? (Hint: Water's specific heat capacity is 4,184 J/kg•°C.)

    ANS:
    Energy transferred = 4,184 J/kg•°C × 1.3 kg × 80°C = 435,136 J

    DIF: II   OBJ: 6-2.4

Examine the graph below, and answer the questions that follow.

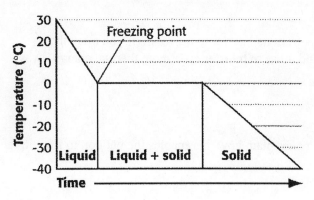

34. What physical change does this graph illustrate?

    ANS:
    The physical change illustrated in this graph is freezing, a change of state from liquid to a solid.

    DIF: II   OBJ: 6-3.2

35. What is the freezing point of this liquid?

    ANS:
    The freezing point of this liquid is 0°C.

    DIF: II   OBJ: 6-3.2

36. What is happening at the point where the line is horizontal?

    ANS:
    A change of state; energy is being transferred away from the substance and the attraction between particles is increasing.

    DIF: II   OBJ: 6-3.2

37. What happens to the particles of a substance as its temperature is increased?

    ANS:
    When the temperature of a substance is increased, its particles gain kinetic energy and, in most cases, move farther apart. The substance will usually expand as its kinetic energy increases.

    DIF: I   OBJ: 6-2.3

38. How are the climates of coastal regions affected by the specific heat capacity of water?

    ANS:
    Water has a high specific heat capacity, which means that its temperature changes slowly. As inland temperatures drop in the winter, the ocean keeps coastal areas moderately warm by retaining thermal energy. In the summer, when inland temperatures are greater, the ocean helps to cool coastal areas.

    DIF: I   OBJ: 6-2.2

39. Which temperature is the warmest: 130°C, 180°F, or 373 K? Show your work.
    (Hint: $°C = \frac{5}{9} \times (°F - 32)$; $K = °C + 273$)

    ANS:
    180°F: 5/9 × (180°F − 32) = 82°C
    373 K: 373 K − 273 = 100°C
    130°C is the warmest of the temperatures given. Students may arrive at this answer by converting given temperatures to Fahrenheit, Celsius, or Kelvin.

    DIF: II   OBJ: 6-1.3

40. In the past, people often used hot objects to heat their beds. Which would keep a person warmer throughout the night: a 1,000 g iron bar or a 1,000 g bottle of water heated to the same temperature? Explain.

    ANS:
    Iron heats and cools quickly because of its low specific heat capacity. Water takes a long time to heat and to cool because it has a high specific heat capacity. Therefore, the heated water-filled bottle would be a better choice than an iron bar of equal mass and temperature.

    DIF: II   OBJ: 6-2.3

41. Explain why the air pressure in an automobile tire is greater after the car has been driven for a few miles than when the car has been parked for several hours.

    ANS:
    Friction between the tires and air inside them increases when the car is driven. As a result, there are more collisions between the air particles in the tire and the air increases in temperature. Air expands when its temperature increases; therefore, the air pressure in the tire increases.

    DIF: II   OBJ: 6-2.4

42. The graph below shows the changes of state that occur as energy is continuously added to 1 g of water. Use the graph to answer the following question.

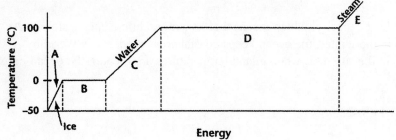

In which regions of the graph is the average kinetic energy of water particles increasing? Explain.

ANS:
Temperature increases at parts **A** (ice), **C** (water), and **E** (steam); therefore, the average kinetic energy of water molecules increases in these regions of the graph.

DIF: II   OBJ: 6-3.2

43. Use the following terms to complete the concept map below: *density, convection, greenhouse effect, radiation, conduction, metal.*

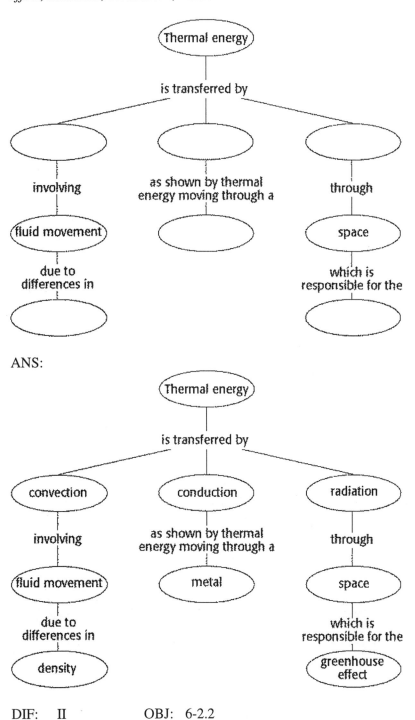

ANS:

DIF: II   OBJ: 6-2.2

44. Why must scientists *average* the kinetic energy of a substance?

ANS:
Particles of matter are constantly moving, but they don't all move at the same speed and in the same direction all the time.

DIF: I   OBJ: 6-1.1